"This book makes a strong case that opposing nuclear energy and driving up its cost turn out to have been historic self-defeats by the environmental movement given the importance now attached to emissions."

— Matt Ridley, author *How Innovation Works*

"Nobody has done more to expose the anti-human roots of the anti-nuclear movement than Robert Zubrin. By weaving together little-known facts with crucial historical episodes, Zubrin has made an essential contribution to our understanding of the war on nuclear. *The Case for Nukes* is a must-read for anyone who cares about the future of the planet and the future of humanity."

— Michael Shellenberger, best-selling author of *Apocalypse Never* and *San Fransicko*

"*The Case for Nukes* is absolutely a work of genius. It takes the vast complexities of the workings of and history behind nuclear energy and explains it all in a most interesting, page-turning manner. It is rare when I learn something new on every page, but I did in Zubrin's *The Case for Nukes*. This Providence-given energy source can allow our civilization to not only survive but go on to a wonderful and most marvelous expansion while maintaining our beautiful planet. I thoroughly endorse this book."

— Homer Hickam, Author *Rocket Boys/October Sky*

"*The Case for Nukes* is a terrific book. Zubrin pulls no punches showing that we already have the technology to provide human civilization with unlimited and clean energy."

— Marian L. Tupy, Editor *Human Progress*

"In *The Case for Nukes* the always provocative Robert Zubrin makes a strong case that technological innovation can solve the world's most crucial and daunting dilemmas."

— Clifford D. May, founder and president,
Foundation for Defense of Democracies

"Robert Zubrin's exposition of the history and science of nuclear power is fascinating. His account of the malicious and sustained campaign of disinformation, distortions, and fear that has denied humanity nuclear power's enormous benefits is infuriating. This book needs to be sent to every American and European old enough to vote."

— Claire Berlinski, Editor *The Cosmopolitan Globalist*

"*The Case for Nukes* is the best book I've read on how to harness the incredible promise of nuclear energy. Robert Zubrin clearly explains why the skyrocketing costs of nuclear are unnecessary—the product of crippling, irrational regulations imposed by badly-motivated environmental activists. Most importantly, he offers a clear blueprint for liberating nuclear so that it can provide low, cost-reliable energy for billions of people for centuries and millennia to come."

— Alex Epstein,
Author of *Fossil Future* and *The Moral Case for Fossil Fuels*

*To the Prometheans*
*Who do not steal fire from heaven*
*But who seek the heavenly fire that is our birthright*
*To break our chains*
*Forever.*

**ALSO BY ROBERT ZUBRIN**

*Islands in the Sky*
*The Case for Mars*
*Entering Space*
*First Landing*
*Mars on Earth*
*The Holy Land*
*Benedict Arnold*
*Energy Victory*
*How to Live on Mars*
*Merchants of Despair*
*Eleanor's Crusades*
*Mars Direct*
*The Case for Space*

# The Case for Nukes

## How We Can Beat Global Warming and Create a Free, Open, and Magnificent Future

## Robert Zubrin

POLARIS BOOKS

*The Case for Nukes:*
*How We Can Beat Global Warming*
*and Create a Free, Open, and Magnificent Future*

Polaris Books
11111 W. 8th Ave. Unit A
Lakewood, CO 80215

www.polarisbooks.org

ISBN 978-1-7363860-6-4  paperback edition
ISBN 978-1-7363860-7-1  e-book

Layout and cover by Marie Stirk

# TABLE OF CONTENTS

# INTRODUCTION

*I come among you not to make peace, but with a sword.*
— Mathew 10:34

**THIS IS A** heretical book. If you are a person who needs to fit in with any established political tribe, it will make you uncomfortable. That, dear reader, I must warn you, is my intention. Party line opinions put reason to sleep, and, as recent history has amply shown, the sleep of reason produces monsters. So I intend to disturb as many people as I can.

Now that you have been fairly warned, let's get right to it. Here's the truth:

1. Global warming and anthropogenic atmospheric chemistry change are both real.
2. They are *not* currently a crisis.
3. But they are going to become a crisis, and then a disaster, unless something is done to effectively change the current trajectory of events.
4. The primary solution offered by those who recognize this problem — to wit, reducing carbon use by making fuel less

affordable to people of limited means — is unethical and impractical, and consequently deserves to fail, has failed, and will inevitably continue to fail, spectacularly.

5. That the claim that modern civilization can be powered by updated forms of the renewable energy sources that needed to be replaced by fossil fuels to enable the birth of industrial society is nonsense.

6. That the more radical prescription of global population reduction offered by the minority of climate crisis believers who recognize the unfeasibility of the carbon tax and green energy solutions is the worst idea of all, one that would lead to catastrophes too horrific to even contemplate were its enforcement seriously attempted.

7. That far from contracting our energy use, human progress must and will inevitably entail continued exponential growth of human power generation.

8. That therefore the widespread adoption of nuclear energy is essential for a positive human future.

I understand that few readers will agree with more than five of the above eight points. But if you are so exceptionally open-minded as to still be reading this, I ask you to hear me out. Those with great gifts bear great responsibilities, and the ability to think is the greatest gift of all. So really, it's your duty — your moral responsibility as a rare free-thinking citizen — to read this book. If you put it down now your conscience may haunt you for the rest of your life. Why risk it, particularly when reading something as original, enlightening, and interesting as this book is going to be so much fun?

Give me a chance, because in this book I'm going to prove all eight points, and then some. I'm going explain how nuclear power works, how much it has to offer humanity, and debunk the toxic falsehoods that have been spread around to dissuade us from using it by variously the ignorant, the fearful, the fanatical, and by cynical political operatives bought and paid for by competing interests. I'm going to tell you about revolutionary developments in the field, including new reactor types that can be cheaply mass produced, that cannot be made to melt down no matter how hard their operators try, that use a new fuel called thorium far more plentiful than uranium, and still more advanced systems, employing thermonuclear fusion – the power that lights the sun – to extract more energy from a gallon of water than can be obtained from 300 gallons of gasoline. I'm going to tell you about the bold entrepreneurs – a totally different breed from the government officials who created the existing types of nuclear reactors – who are leading this revolution in power technology. I'm going to make clear the critical difference between practical environmentalism, which seeks to improve the environment for the benefit of humanity, and ideological environmentalism, which seeks to use instances of human insult to natural environment as evidence for a prosecutorial case against human liberty. I'm going to show how the latter school of thought is wrong, not only with respect to the catastrophic harm it would do to humanity, but to nature as well. And, oh yes, I will also tell you about mercenary environmentalism, which seeks to deploy troops of dupes to shut down companies or whole industries in order to eliminate competition and be suitably rewarded by the beneficiaries of such efforts in return. That's a very important environmental movement, too, whose omission from any discussion of energy policy would be completely unfair. I'm going to show that, when it comes to environmental improvement, freedom is not the problem; freedom is the solution. I'm going to make clear both the possibility and necessity of the nuclear revolution by putting it in a broader historical context. I will explain the relationship of nuclear energy to the overall process of development of civilization, whereby new technologies create new resources and new knowledge, which in turn make possible still more technological advance.

Finally, I'm going to bring all this to bear on addressing the greatest threat facing humanity today – which is the possibility that we will turn on each other, as we did in the 20th century, under the spell of the false idea that resources are finite.

Only in a world of unlimited resources can all men and women be brothers and sisters. Only in a world of freedom can resources be unlimited.

That is the world we can, and must, create.

This book will explain how.

# CHAPTER 1

# TOO MUCH SMOKE, TOO LITTLE FIRE

**THE WORLD CURRENTLY** faces two energy crises: We have too little energy, and we have too much.

We have too little energy, because the main problem that the bulk of humanity faces today, every day, is poverty. To provide a decent standard of living for all, humanity is going to have to generate and put to use many times more energy than it does today.

We have too much energy, because at the rate we are currently using it, we are measurably changing the Earth's climate and chemistry, and if we keep increasing our use — which we must and will — we could change it in ways that prove catastrophic on a global scale.

Many drawn from the more well-off part of the population in the world's advanced sector have chosen to focus on the second problem, proposing to reduce the use of fossil fuels by taxing it, thereby making such necessities less affordable to people of limited means. I believe that such approaches to the problem are unethical, and while people may debate their ethics, there is no debating the fact that they have not worked. Indeed, they have failed spectacularly. Between 1990, when world leaders first mobilized to try to suppress $CO_2$ emissions, to today, total global annual carbon use doubled from 5 billion tons to 10 billion tons. This followed a pattern of doubling our carbon use every thirty years for more than a century. In 1900, humanity burned 0.6 billion tons of carbon per year. This doubled to 1.2 billion in 1930, doubling again to 2.5 billion tons in 1960, then yet again to 5 billion tons in 1990, to 10 billion tons now.[1]

The reason for this increase is simple. Energy is fundamental to the production and delivery of all goods and services. If you have access to energy, and the things made by energy you are rich. If not, you are poor. People don't like being poor, and they will do what it takes to remedy that condition. Despite the Depression, two world wars, and all sorts of other natural and human-caused catastrophes over the recent past, they have, on the whole, been very successful at finding such remedies. In 1930, the average global GDP per capita, in today's money, was $1500/year. Today it is about $12,000/year, an increase closely mirroring the climb in energy consumption. This rise in energy use has enabled a miraculous uplifting of the human condition, dramatically increasing health, life expectancy, personal safety, literacy, mobility, liberty and every other positive metric of human existence nearly everywhere. But $12,000/year is still too low. In the USA we average $60,000/year, and there is still plenty of poverty here. To raise the whole world to current American standards will require multiplying global energy use at least fivefold – and probably more like tenfold once population growth is taken into account.

Based on history, there is every reason to expect human energy production and use to double again by 2050, and yet again by 2080. Every person of goodwill should earnestly hope for such an outcome, because it implies a radical and necessary improvement in the quality of life for billions of people. That in fact is why it is going to happen, whether ivory tower theorists wish it or not. Humanity is not going to settle for less.

But human energy production today is overwhelmingly based on fossil fuels. Were it to remain so, while we double and redouble our energy use, the effects on the planet would become serious.

Global warming is a fact. It is unfortunate that the debate over it has been politicized to the point where opposing partisans have chosen to either deny it or grossly exaggerate it. Neither approach is helpful. So, I'm not going to indulge in the customary hysteria and tell you that we have only 18 months or 18 years to decarbonize the economy or face doom by fires and floods, catastrophic rainfalls and droughts, evaporating poles or glaciers advancing rapidly together with unstoppable armies of ravenous wolves, or similar biblical plagues held by some to be responsible for the Great Silence of the millions of extraterrestrial civilizations driven to their extinction by their inability to pass the Great Filter of global warming.[2]

Nevertheless, global warming is certainly real. According to solid measurements, average temperatures have increased by about 1 degree centigrade since 1870. That, admittedly, is not a big deal. It is the equivalent to the warming that a New Yorker would experience if he or she moved to central New Jersey. So there is no climate catastrophe *now*. But the climatic effects of continued $CO_2$ emissions at a level an order of magnitude higher than today would be an entirely different matter.

Moreover, while climate change will still take some time before it becomes an acute matter (in reality, as opposed to agitation), other effects of the emissions of fossil power plants are already quite serious. First among these is good old-fashioned air pollution, *not* $CO_2$, but particulates, carbon monoxide, nitric oxides, and sulfur dioxide. Worldwide these emissions are currently killing people at a rate of over 8 million per year,[3] and causing many billions of dollars of increased health care costs. Then there is the $CO_2$ itself. While $CO_2$-induced warming has only raised global temperatures by about 1 percent of the

100-degree Centigrade range inhabited by terrestrial surface life from the equator to the poles (or 0.35% of the 287 K average absolute temperature of the Earth's surface) since 1870, the combustion of fossil fuels has raised the atmosphere's $CO_2$ content by *50 percent* (from 280 ppm to 420 ppm). That's a lot. A fifty percent increase in $CO_2$ represents a truly significant change in the Earth's atmospheric chemistry, with readily observable effects. In some respects, these effects have actually been beneficial. For example, as a result of $CO_2$ enrichment of the atmosphere, the rate of plant growth *on land* has increased significantly. There is no doubt about this. NASA photographs taken from orbit show an increase in the average rate of plant growth of 15 percent worldwide since 1985.[4] That's great, but there is a problem: we have seen no comparable improvement in the oceans. Quite the contrary. Evidence is mounting that increased acidification of the ocean caused by takeup of $CO_2$ is killing coral reefs and other important types of marine life. There are things we can do to counteract this, for example farming the oceans, as I will discuss, to put some of this excess $CO_2$ to work to increase the abundance of the marine biosphere. But there are limits to the capacity of marine and even terrestrial biomes to take advantage of increased $CO_2$ fertilization of the atmosphere. If fossil fuel use continues to rise exponentially to support world development, this capacity will be overrun. All fertilizers – such as nitrates, or even water – become harmful when present in too great abundance. This could well become the case should our current river of $CO_2$ emissions become a flood. It is not certain at what point the biosphere's defenses will fail, but that is not an experiment we should not wish to run.

In saying all this, I do not wish to make a case against fossil fuels. The emissions resulting from burning such fuels may be killing 8 million people every year, but the energy they produce is enriching the lives of billions. The positive transformation of human life that has been accomplished by the massive increase in power enabled by fossil fuel use is beyond reckoning, and far from abandoning it, we must – and will – take it much further.

So, the bottom line in this: we are going to need to produce a lot more energy, and it will need to be carbon-free. The only way to do that is with nuclear power.

# CHAPTER 2

# A BRIEF HISTORY OF POWER

**ANCIENT LEGENDS TYPICALLY** ascribe the rise of humanity to one of two gifts bestowed by, or stolen from, the gods: fire and speech. These are indeed great possessions – fundamental attributes without which human progress, or even existence above the condition of animals, would not be possible. In fact, our distant ancestors had fire before they were able to speak. It was their mastery of this key technology that allowed them to become human.

Simplifying greatly for the sake of brevity, the ancestral history of humanity may be described as follows. Over the period from 10 million to 4 million years ago, a group of apes descended from the trees to take up life on the African savannah. These hominids, as they are called, developed a number of human features, including upright posture and opposable thumbs. This opened the way to the development of the ability to throw projectiles, humanity's one unique point of athletic superiority among animals. This talent enhanced the hominids' ability to hunt and defend themselves and put a premium on the development of certain parts of the brain relating to these activities. These creatures then gave rise to the first human species, known as *homo habilis*, or "clever man," about 2.3 million years ago. Homo habilis were little people, about one meter tall, with brains 650 cc in size. This is less than half that of modern humans, but nearly double that of chimpanzees. So endowed, Homo habilis were able to fashion simple stone tools known as hand axes, increase their numbers, and successfully spread over a considerable part of Africa.

They did not, however, have fire. So, while homo habilis could hunt, and cut meat with their hand axes, they had to devote a considerable part of their metabolic energy to digesting the meat they managed to acquire. This limited the amount of energy available for their brains, and since brains require a lot of energy, this limited their brain size and potential mental development.

But fire was intermittently present all around them, and at some point, a member of this species hit on the idea of putting it to use. Undoubtedly, they knew it could keep them warm. Many animals like to sleep next to a fireplace or campfire. Transporting a burning stick to start and then feed and manage a controlled fire to obtain heat on demand, however, is a technology – a practical application of an understanding of a phenomenon – something no other animal had ever done before. And once campfires became a part of daily or nightly life, it was only a matter of time that some meat fell into the fire, and people discovered that it tasted better cooked. That changed everything.

By cooking their food, people could eat more meat, far more safely, and save their metabolic energy for their brains, instead of their stomachs. This change in the human condition led very quickly about 2 million years ago to the evolution of a new species of humans, *Homo Erectus*, with greater stature and a brain size fifty percent larger than homo habilis. Armed with fire, a better physique, and more smarts, Homo Erectus was able to spread beyond tropical Africa to Europe and Asia. This much larger and more diverse population then gave birth a few hundred thousand years ago to our species of human, *Homo Sapiens* – a species smart enough to invent language.

By inventing fire, humanity invented itself.

Homo Erectus knew how to cook meat and warm themselves with fire. Brainier Homo Sapiens figured out a lot more things they could do with it, including smoking meat and fish for preservation, baking grains into breads, making hides for clothing and tents, transforming wood into a superior fuel in the form of charcoal, and creating pottery. With the help of language, first spoken, later written, as well as art and other forms of symbolic communication, such innovations could be transmitted from place to place in the form of ideas. As inventions made anywhere could now spread everywhere, progress was accelerated, and pottery kilns led to brick kilns, and charcoal-fired kilns led the way to metals and glass.

Smoked meats, breads, leather, charcoal, pottery, bricks, glass, and metals are all artificial materials. They do not exist in nature.

By inventing fire, humanity created the material basis of its world.

In short, fire, the ability to release concentrated energy to effect chemical changes in matter, is fundamental to human existence.

Fire requires fuel. The first fuel employed was wood. But wood contains tars, which produce a smoky flame, that can be very unpleasant if released into a confined environment like a cave or a hut. As early as Neolithic times, however, people found they could reduce this problem by baking the tars out of the wood in a hot pit outside their huts, before they used it to cook inside. Thus was born charcoal, the first artificial fuel. Other artificial fuels followed, including olive oil, according to legend the gift of Athena to her namesake city, which in its heyday produced and exported oil on an industrial scale to light the lamps of much of the classical Mediterranean world.

As civilization progressed, it became ever more energy intensive. Starting in the Bronze Age, wind power was harnessed, enabling long-distance maritime transport of massive amounts of commercial goods. In classical times, water wheels were introduced, allowing the power of moving rivers and streams to be used to do all kinds of mechanical work, replacing mind-dulling brute work. This was particularly liberating for women and girls, who previously were condemned to spend much of their mornings grinding grain to make bread. The Greek poet Antipater of Thessalonica (circa 10 AD) hailed the amazing power of the new invention.

"Hold back your hand from the mill, you grinding girls," he wrote. "Even if the cockcrow heralds the dawn, sleep on. For Demeter has imposed the labors of your hands on the nymphs, who leaping down upon the topmost part of the wheel, rotate its axle; with encircling cogs, it turns the hollow weight of the Nisyrian millstones. If we learn to feast toil-free on the fruits of the earth, we taste again the golden age."[2]

**Fig. 2.1** The Grand Pont in Paris. To exploit all available water power, the spaces under bridges were filled with water wheels.

**Fig. 2.2** Depiction of women using a water mill to grind grain, taken from the Twelfth Century manuscript of Herrade de Landsberg. Note the complex gearing. Powered mills had the same significance for women of the Twelfth Century as washing machines did for those of the Twentieth.

Water wheels became incredibly popular. The Norman Domesday book lists over *ten thousand* of them operating in newly conquered Saxon England. They were used not only to grind grain, but to cut wood, marble, and stone, drive windlass cranes to construct castles and cathedrals, and to pump water. By the high Middle Ages, as the mechanization of labor became a defining feature of Western civilization, every bridge in Paris, and probably most of France, hosted dozens of water wheels clanking away, day and night.[3]

River power was limited by geography and was so aggressively exploited that it threatened to be tapped out. But in the 12th century, French mill tinkerers rose to the occasion by inventing the post windmill. This clever device solved the

**Fig. 2.3** The post windmill could turn to face the wind. Invented in northern France during the mid-twelfth century, it radically expanded Europe's mechanical energy resources and contributed to rapid economic growth.

difficult problem of tapping the power of a moving fluid coming from an unknown direction by using the force of the wind itself to *automatically* turn the mill to face the right way. Now mechanical power could be available everywhere!

Women freed from grinding grain could spend more time raising children, gardening vegetables, taking care of animals, sewing, or weaving clothes, all of which contributed to a rising population and living standards. Moreover, as ever more brute labor tasks became mechanized, the utility of slavery in the West declined, and its role in the economy and society became ever more marginal.

In contrast, in non-Western societies which failed to embrace mechanized power, commoners remained in a much more degraded state, with the worst example being sub-Saharan Africa. There, because of the difficulty of maintaining draft animals, tasks done by beasts in Europe and Asia were forced upon

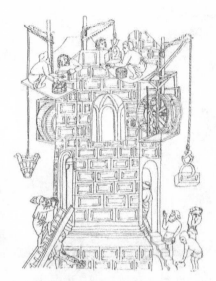

**Fig. 2.4** In Medieval Europe, construction was done using complex machinery

slaves, who therefore became a principal commodity for export once these regions were brought into contact with the Islamic and European worlds.

Technology is the foundation for freedom.

As the European population multiplied and living standards rose, their fuel needs could no longer be met by wood or other biofuels, whose annual rate of production is limited by the diffuse flux of sunlight on the land, transformed into chemical energy at about 1 percent efficiency by plants. By the 1500s, these limits began to threaten parts of Europe with deforestation. But fortunately, the same vibrant European Renaissance civilization whose growing dynamism appeared to threaten the continent's forests instead used its creativity to save them by inventing a way to power itself with the stored energy of sunlight from ages past.

Trees alive today need their carbon. Those that died hundreds of millions of years ago do not. Fossil fuels are the material remains of plants and animals long dead, locked away for ages from use by the active biosphere. By putting first coal, and then oil and gas, to use, humans sent these frozen assets back into circulation. Then, having created such vastly expanded raw energy resources, we found extraordinary new ways to set them to work in first steam and then internal combustion, diesel, and gas turbine engines enabling unprecedented mobility by land, sea or air, or transmissible in the form of electricity to instantly provide mechanical work, heat, or light, anywhere, anytime, on demand. Furthermore, in the process of freeing ourselves from the limits imposed by depending on nature for such supplies, we lifted the pressure we previously had been placing on diverse living energy sources, ranging from forests to whales.

This is an important point that deserves emphasis. There are people today who believe they can live most "sustainably" by living "naturally," that is, depending upon nature for their sustenance. They could not be more wrong. Whales do not want you to depend on them for your fuel, and if trees could think, neither would they. Humans would not benefit the wild biosphere by relying on it for our consumption. On the contrary, we would destroy it. But we can preserve wild nature – and in fact, as I will discuss later, greatly enhance it – by creating and relying on resources that are *not* part of it.

To preserve the natural, we must create the artificial.[4]

This brings us to the subject of the human situation today. Modern civilization relies on fossil fuels. That is a good thing. By creating an enormous energy base for ourselves through the production of vast amounts of cheap fossil fuels, we have enabled a magnificent global civilization of eight billion people with health, life expectancy, prosperity, opportunity, literacy, leisure, mobility, dignity, liberty, and benefits in every other category of existence far exceeding the wildest dreams of any previous age. That's the truth. Despite all the problems, conflicts, limitations, and very real defects of modern society, people have never, remotely, been as well off as they are today – and this has only been possible because of a terrifically powerful world economy driven by fossil fuels. Furthermore, we have done this while preserving enough of wild nature to provide endless aesthetic enjoyment to anyone with the inclination to partake.

True enough, all but those blinded by some sort of fanaticism must admit. But can this go on? We are using energy at a fantastic, and rapidly increasing, rate. Won't it run out? And what of the side effects? All this fossil fuel use is changing the planet's atmosphere, adding carbon dioxide, a gas that traps heat like the walls of a greenhouse and affects the chemistry of the ocean. Aren't we inviting a $CO_2$ catastrophe?

So, in brief, *can we really keep this up*, and *what happens if we do*? Those are very important questions. For the rest of this chapter, I will devote myself primarily to answering the first. The second, bearing on global warming, air pollution and ocean acidification, deserves an ample discussion all to itself, which I'll deliver in Chapter 14. But in any discussion of energy policy, it can't be set aside entirely. So, at this point I'll state my view, to be justified at length later, that fossil fuel combustion-driven changes in the Earth's climate and chemistry are certainly real. Furthermore, they are most probably, to a significant degree, human caused. That said, it is a *problem*, not a *crisis*. The distinction is important.

We all face many problems in our lives that we need to continually solve on an assortment of different time scales. We need to breath every minute, drink water every day, eat at least every several days, pay our rent or mortgage every month, our taxes every year, and plan for our retirement over decades. We need to deal with all of this. But if you are a SCUBA diver deep underwater and your air is running out, that is the crisis you need to deal with. Your rent payment or retirement planning problems can be dealt with later. That said, medium-term considerations can put boundary conditions on what short-term actions might be acceptable. For example, it would not be wise if you were the diver mentioned above to try to solve your problem by murdering another diver and stealing his air tank, because that could get you into a lot of trouble shortly after you surface. But while you need to keep such unviable alternatives out of bounds, it remains the case that the problem you need to focus on is your air shortage, not your rent.

The immediate energy problem is to continue to provide enough power for a growing and ever-advancing human civilization. The global warming, air pollution, and ocean acidification problems will need to be addressed eventually, and so we should avoid solutions to the supply problem that will make that harder. But it is simply not true that we only have two years, or ten years, or twenty years, to stop our carbon emissions, or that radical regressive measures to try to force things in that direction by making fuel unaffordable are appropriate. Quite the contrary. We have decades to take the required measures to mitigate global warming, and we will only be able to develop them, and afford them, if we remain free and prosperous.

Freedom is not the problem. Freedom is the solution. Prosperity is not the problem. Prosperity is the solution.

So how can we use our present freedom and prosperity to create an energy foundation for more of both in the future?

We begin by discussing the world as it is.

## FOSSIL FUELS

Fossil fuels are the basis of modern civilization. They provide close to 100 percent of all energy supporting transportation,

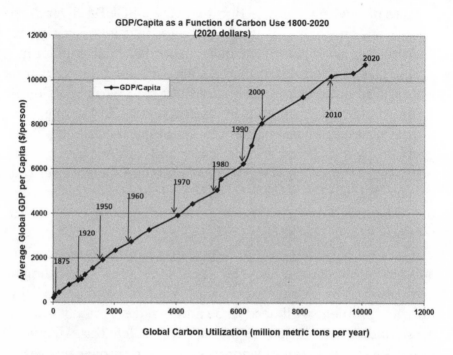

**Fig. 2.5** Human living standards have improved in direct proportion to worldwide fossil fuel use.[5]

industry, and agriculture, and nearly 90 percent of domestic energy use. Those people who claim we can decarbonize our lives by taxing gasoline 10 percent, 100 percent, or 1000 percent, for that matter, are talking nonsense, because everything that everyone eats, wears, lives in, or otherwise uses, was both produced and delivered to them through the utilization of fossil fuels. The benefits of doing so have been profound, as shown in the graph ~~below~~ *above*, which compares average gross domestic product (GDP) per capita to total worldwide fossil fuel use.

It can be seen that between 1800, when the industrial revolution commenced, and the present, human living standards, measured in constant dollar GDP/capita, have improved in direct proportion to our total fossil fuel use. Furthermore, since the size of the human population has increased at roughly the same rate, what the data shows is that the total world GDP has increased roughly in proportion to the rate of fuel use *squared*. In secular terms, the tale told in the graph is the greatest story ever told. It is the rise of billions of people out of poverty, accomplished through the liberal use of energy. Rolling this back would be catastrophic, and almost certainly genocidal. Rather than attempting any such misconceived project, the challenge presented by the tale of the graph is how to continue it. Through the use of fossil fuels we have raised global living standards from around $200/year per capita in 1800 to over $12,000/year per capita today. That's amazing, but hardly good enough. The average GDP/capita in the USA is around $60,000, yet we still have too much poverty here. If we are to achieve a world where everyone can have a decent life, we are going to have to use *a lot more energy*, not less.

While the achievements of fossil fuel technology are undeniable, our dependence on this resource raises issues. *We are using so much. Won't the stuff run out?*

These concerns are not new. In 1865, British economist William Stanley Jevons published a book called *The Coal Question*, in which he warned that with their rising use, coal supplies must run out in a few decades, aborting Britain's industrial revolution. "Are we wise," he asked, "in allowing the commerce of this country to rise beyond the point at which we can long maintain it?"

Jevons' prophesy of imminent fuel exhaustion proved to be nonsense, but that did not stop a string of emulators from following in his footsteps. For example, in 1874, the state geologist of Pennsylvania, then the world's leading oil producer, estimated that the USA had only enough oil for another four years. In 1914, the Federal Bureau of Mines said we had only ten years of oil left. In 1940, the bureau revised its previous forecast and predicted that all our oil would be exhausted by 1954. In 1972 the prestigious Club of Rome, utilizing an inscrutable but allegedly infallible MIT computer oracle, handed down the ironclad prediction that the world's oil would run out by 1990.[6] The Club said at that time that only 550 billion barrels were left for humanity. Since then, we have used over a trillion barrels, and are now looking at proven reserves of over 1.7 trillion more. Since 1972, there have been repeated predictions of imminent oil supply exhaustion published every few years by various authorities, and not one has come true. In fact, if we look at the ratio of proven reserves to consumption rate, the world today has a bigger oil supply now than it ever has at any time in the past!

Gas supplies are increasing too. In fact, despite a 50 percent increase of US natural gas consumption over the past two decades (or rather *because* of it) American gas reserves have tripled since 2000. This can be seen in the data in Fig. 2.6, published by the US Energy Information Administration. A similar pattern can be seen with the growth of US oil reserves as well.

How is this possible? Surely if we are consuming our oil and gas, our reserves should be shrinking, not growing. Yet growing they are, as they have been since 1859, when there were zero known US oil reserves.

The answer is new technology. It is technology that defines what is and is not a resource, of any kind, emphatically including oil and gas reserves. In the case at hand the new technology is hydraulic fracturing, commonly known as fracking.

Frackers employ horizontal drilling technology to send drills sideways in several directions from a single vertical hole, and then expand access to oil and gas trapped between neighboring layers of shale by pumping water down at very high pressure. The high-pressure water fractures the shale, creating openings for fluid flow that are kept open by sending sand or other gritty particles down with the water. While radically expanding our

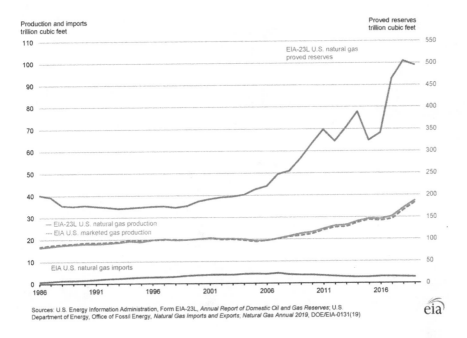

**Fig. 2.6** Despite expanded use of natural gas, US gas reserves are continually increasing.[7]

oil and gas resources, fracking greatly reduces the environmental footprint of the industry. In a conventional oilfield you may see scores of pumping derricks per square mile, each drawing oil from a narrow region below. In a fracked field, you are likely to see just one, because that is all that it takes to access a large underground area.

This technology was developed by independent Texas oilmen in the late 1990s, and began to yield important results circa 2006. In 1990, North Dakota had no significant oil reserves or production. Now it has vast reserves, and it is the third largest oil and gas producer of any American state.

Europe has been an oil importer since the beginning of the petroleum era. In fact, it was the Germans' lack of oil supplies

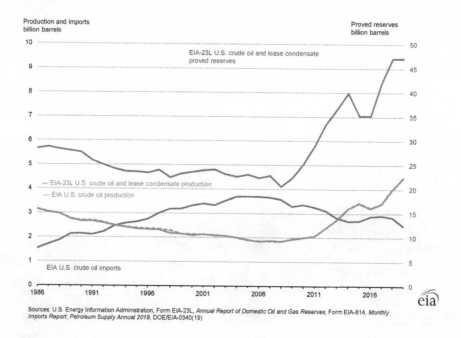

Production and imports
billion barrels

Proved reserves
billion barrels

EIA-23L U.S. crude oil and lease condensate
proved reserves

— EIA-23L U.S. crude oil and lease condensate production
--- EIA U.S. crude oil production

EIA U.S. crude oil imports

1986    1991    1996    2001    2006    2011    2016

Sources: U.S. Energy Information Administration, Form EIA-23L, *Annual Report of Domestic Oil and Gas Reserves*; Form EIA-814, *Monthly Imports Report*; *Petroleum Supply Annual 2019*, DOE/EIA-0340(19)

**Fig. 2.7** American oil reserves are rapidly growing too.[8]

that was a critical weakness leading to their defeat in both world wars.[9] That weakness has been only partially remedied by the development of offshore drilling technology, and with it the North Sea oilfields, since the 1980s. Today Europe remains critically dependent upon Russian gas and Arab oil. But it doesn't have to be that way. There are vast shale oil and gas reserves beneath Britain, France, Germany, Poland, and Ukraine that could transform those countries into fossil fuel exporters, and Europe collectively into a fossil fuel superpower, if only Europeans would choose to embrace fracking technology. The Kremlin, whose number one source of income is gas sales to Europe, has funded a massive Green movement campaign to deter such a decision, and derail the nuclear alternative as well. So far they have been successful.[10] But the reserves are there, awaiting only the light of reason to arrive in the European body politic to transform them into resources.

China is another oil importer with gigantic shale oil and gas formations waiting to be developed. There are many criticisms that can justly be made of the Chinese leadership, but lack of resolve isn't one that comes to mind. It's a good bet that Mr. Putin won't be able to deter them.

But conventional, offshore, and shale oil and gas reserves are only a tiny fraction of the fossil fuels that can be found on Earth. At the bottom of the ocean there lies vast quantities of a substance known as methane hydrate, a sort of ice in which molecules of methane are locked under high pressure in a crystal network of water molecules. One cubic meter of methane hydrate contains 164 standard cubic meters of natural gas. Worldwide, it is currently estimated that about 500,000 trillion cubic feet (TCF) of methane are locked in hydrates beneath the ocean.[11] For comparison, the entire human race currently uses about 132 TCF per year. At that rate, the *currently known* reserves of methane hydrates would be sufficient to last us 4,000 years, – and there are sure to be a lot more that we don't know about. We do not yet have technology to mine this economically, but people are working hard to create it. If we ever should need it, it will be there.

So we are most definitely not about to run out of fossil fuels anytime soon.

But there is another problem, and that is – you guessed it – combustion emissions. The problem is not the global warming we have seen so far. The world has warmed about 1 C since 1870, which is no big deal. For someone living in the United States, that's about the same amount of climate change as you would experience by moving about 90 miles south – a fate innumerable people have endured without difficulty. More significant has been atmospheric chemistry changes, with a 50 percent increase in the air's $CO_2$ content over the same period.

That's enough to matter. The effects of this change on plant life on land have so far been beneficial, while marine life has experienced some damage. But raising the entire world to a decent living standard will require increasing humanity's energy consumption at least tenfold. Were we do to that using fossil fuels, we would experience warming, and — more importantly atmospheric chemistry changes — of a different order altogether.

So we are going to need to create substantial new energy resources that are not carbon based. There are two possibilities available: We can try to go back to the renewable energy sources that preceded fossil fuels, or we can look to new energy sources that fossil-fueled civilization made possible.

Can we go back? Let's see. Biomass was the primary source of energy for humanity prior to fossil fuels. But today it remains an important source of power only for use by the world's poorest people, as it is the least convenient and most polluting source of energy there is. It is also the energy source with *by far* the most destructive environmental impact. For example, a typical coal, gas, or nuclear power plant meeting the electricity needs of a city of 1 million people might have a power output of 1000 MWe, which requires 3000 MWt of heat to generate. If such a power station were fueled with biomass, it would require about 6.3 million tons of wood per year — the yearly production of 1.5 million acres of southern pine. To provide all of America's electricity in this way would require burning 500 million acres worth of forest production every year. Providing the energy to replace our oil-derived transportation fuel would require burning an additional 380 million acres, with industrial and domestic energy needs requiring 470 million more. That's a total of 1350 million acres, or over 2.1 million square miles of forest burned every year- *roughly half the entire area of the continental USA*. More densely populated countries, such as Europe, China, India, and Japan, would have to burn several times their total territory. The effect of such tree harvesting on wildlife would be devastating.

Clearly, we can't power modern society — let alone the still more energy intensive society of the future — with biomass. That leaves water, wind, and solar power.

**Fig 2.8** These racoons would prefer humans generated power without burning their forest. (Credit: Wikipedia Commons)

Modern forms of these dilute power sources would serve better than biomass, but that is not saying much. To produce enough round the clock power to provide the 1000 MWe needed by a city of 1 million, you would need about 10 square miles of solar panels — and triple that much land to put them on, for a total of around 30 square miles — an area 1.5 times the size of Manhattan. The panels would cost you about 25 billion dollars, and getting hold of that much land near most major cities would cost a lot more, and in many cases require leveling the homes of millions of people. Wind power would require four times that much land. Neither of these power sources would be reliable; to provide power at night or during days of cloudy or windless weather they would have to be backed up to the extent of 100 percent of their capacity by other sources. In contrast, a fossil fuel or nuclear power plant producing 1000 MWe, reliably round the clock, could easily be situated on a small fraction of a square mile.

But even so, could such "renewables" make a substantial contribution towards doing the job? There are useful niche applications for these technologies, but as far as providing the foundation for powering the human future is concerned, one look at fig. 2.5 should tell you the answer. In 1800, the human race, then consisting of 1 billion people, was able to generate a GDP/capita, adjusted for inflation to 2020 dollars, of $200/year, based on aggressive – and very environmentally damaging – use of renewable energy. Today, the world population is 8 billion, with much higher energy use per capita, and vastly higher energy use overall. The idea that we can revert to the type of energy sources that were *already proving inadequate in 1800,* and even maintain modern society, let alone develop it much further to create the kind of world we truly need to have, is just bonkers.

Technology is the application of the understanding of phenomena found in nature. In the modern age, such understanding is known as science, but the process existed long before science became a formal disciple. Once the laws of the phenomena are understood, whole new arrays of technology can be created, which, when fully elaborated, provide the tools to break through to the next level of science. Preindustrial humanity's understanding of the laws of combustion, animal and plant breeding, wind and water dynamics, and metallurgy, among others, eventually gave rise to a society which could penetrate the mysteries of thermodynamics, and turn heat into motion using coal-fired steam engines made of steel. That society, equipping itself with railroads, steamboats, and steam-powered factories, then made the breakthroughs to discover the laws of electricity and chemistry. Humanity was then able to elaborate that new knowledge into telegraphs and telephones, light bulbs, power stations, oil wells and refineries, plastics, aluminum, automobiles, radio, electric appliances, and airplanes. That society, in turn, was able to discover the laws of quantum mechanics and nuclear physics. [12]

As each level of science is reached, whole new dimensions of potential inventions become conceivable, and can combine with each other and those enabled by earlier science to enable yet more inventions. The more inventions there are, the more become possible. The process is open-ended, self-accelerating, and potentially unlimited.[13]

The fruits of quantum mechanics are now all around us, in the form of lasers, semiconductors, silicon-chip based computers, the internet, and indeed, the entire unfolding information age. These, however, all exploit phenomenon that are fundamentally electronic in nature. Nuclear physics involves processes that are a million times more energetic, and potentially a million times more potent. It represents a new universe of capabilities, radically transcending anything offered by pre-industrial, thermodynamic, electrical, chemical, or quantum mechanical technology.

Applications of nuclear physics have so far been modest, being decisive in only one field: weaponry. It would be a tragedy beyond description were that to remain the case. It is as if the only permitted application of modern biology was germ warfare. The same power for unlimited harm can also do unlimited good.

# CHAPTER 3

# DO WE REALLY NEED MORE ENERGY?

**THE MOST IMPORTANT** step in any engineering design process is to define the requirements. It is important to design things right, but it is even more critical to design the right thing. So, in trying to design a power plant for human civilization, the key question is: how much do we need? Do we really need more energy? Perhaps we could get by with less energy use in total, with more for each person, if we just had fewer people. Such is the conclusion of the Malthusians.

In 1798, Thomas Malthus, a professor at the British East India company's East India College, published the first edition of his famous *Essay on the Principle of Population.* In this work he argued that human reproduction would always outrun available resources, and consequently the poverty and oppression of not only his employer's subjects in the Indian subcontinent, but of the poor worldwide, was both inevitable and irremediable.[1]

The Malthusian theory has been used as the scientific justification for brutal policies from Malthus' own time to the present. But is it true? On the surface, the idea that the more people there are, the less there will be to go around appears to make sense. It follows that if we get rid of some people (especially those we don't like anyway), we'll all be better off. Thus, those

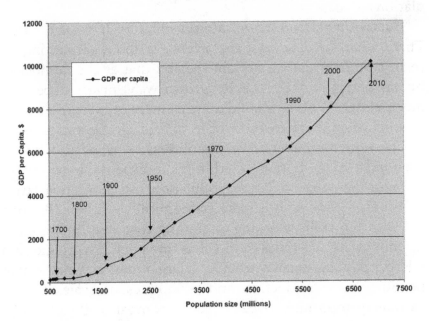

**Fig. 3.1.** How per capita GDP has changed as population has grown, 1500-2010. (2020 dollars)[3]

interested in eliminating Indians, Irish, Jews, Slavs, Africans, or whomever, have been able to argue that their policies, while harsh, are simply necessary to "make the world a better place."[2]

This theory, however, is barking mad nuts. If you doubt that assertion, have a look at Fig. 3.1 which shows global average GDP per capita as a function of population size.

There certainly seems to be a pattern here, but it is obvi-ously *not* the Malthusian claim that living standards decrease as population grows. Rather, what we see is GDP per capita *increasing* with population. From 1800 to 2010, world popu-lation increased sevenfold while the average GDP/capita rose more than fifty times. So, the GDP/capita has risen as popula-tion squared, while the *total* GDP has risen, not in proportion to the population size, but in proportion to the size of the population *cubed!*

"But that makes no sense!" Malthusians cry. It doesn't matter. That's what the data says, and science is about accounting for reality. So how can we explain the *fact* that as the number of human beings on the planet has grown, we've nearly all become much better off? Why should there be more of everything to go around, when there are more of us to feed, clothe, and house?

There are a number of very good reasons why this is so. As economist Julian Simon noted in his indispensable book, *The Ultimate Resource,*[4] a larger population can support a larger division of labor, so it is more economically efficient. Ten people with ten skills, working or trading together, can produce far more than ten times as much as one person with one skill. A larger population also provides a larger market which makes possible mass production and economies of scale. This is extremely important, as we can see by comparing the price of a rocket engine with that of a small car. The RL-10, a tried-and-true thruster which has been in production since the 1960s, contains less metal (about 500 pounds) and is significantly less complex than a typical small car. Yet RL-10s sell for around $6 million each, while a new compact car can be obtained for less than $20,000. This is because there is only a market for a few RL-10s per year, while cars are sold by the millions. Because they represent a larger market, larger populations drive investment in new plant and equipment much more forcefully than small populations. If the market for an item is small, no one is going to build a new factory to produce it or spend much money on research to find ways to improve it. But if the sales opportunity is great, the necessary investment will occur instantly as a matter of course. A larger population can much better justify and afford to build transportation infrastructure such as roads, bridges, canals, railroads, seaports, and airports, all of which serve to make the economy far more efficient and productive. A larger population can also better afford to build other kinds of highly productive economic infrastructure, including electrification and irrigation systems. It can better afford infrastructure necessary for public health as well, including hospitals, clean water, and sanitation systems, and act far more effectively in suppressing disease-spreading pests. It takes a large-scale effort to drain a malarial swamp, a reality that puts such projects beyond the capability of small, highly dispersed populations such as those that still exist in many parts of Africa. Furthermore, human boots on the ground are necessary to patrol the regions in which we live to prevent ponds and puddles from being used by mosquitoes and other disease carriers as breeding grounds. A thin population will thus in many cases tend to be a much sicker population than a dense population which enjoys the safety that only numbers can provide against humanity's deadly natural enemies. And again, a healthy population will be more productive than a sick population and reap a much better return on the investment it chooses to make in education (and thus be able to afford more education), since more of its young people will live to employ their education, and be able do so for longer life spans.[4]

That said, it is clear that the primary causative agent for higher living standards is not population size itself, but the overall technological development that it allows. The average living standard is defined by GDP available for consumption per capita, which is equal to the production per capita, which is determined by technological prowess.

If we choose to be mathematical, we could even write this down as an equation. Let $L$ = Living standard, $P$ = Population, $G$ = Gross Domestic Product, and $T$ = Technology. Then we have:

$$L = G/P \qquad \text{Living standard = GDP/Population} \qquad (1)$$

$$G = PT \qquad \text{GDP = Population} \times \qquad (2)$$
$$\text{Technology}$$

Putting equations (1) and (2) together, we find, simply that:

$$L = T \qquad \text{Living standard =} \qquad (3)$$
$$\text{Technology}$$

So, the question is, what causes the advance of technology? Well, clearly, technology does not come from the land, it comes from *people*. It is the product of *human work*. The most general way to measure human work is in terms of person-years. So, let's graph the growth technology (measured in dollars as GDP per capita) against person-years expended, from the year 1 AD to the present. The results are shown in figs 3.2 and 3.3. (I've

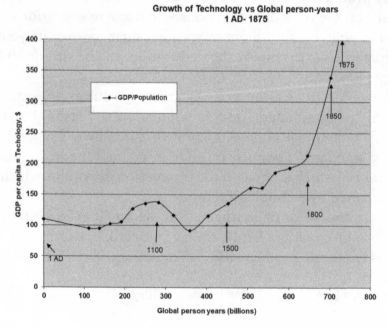

**Fig. 3.2** Growth of Technology with respect to person-years, 1 AD to 1875[5]

used two graphs to show this to avoid the necessity of using logarithmic scales, which are harder to read.)

Let's look first at Fig. 3.2, which shows the growth of human technology worldwide from the time of the Roman and Han Empires to the late nineteenth century. There's a lot of very interesting history to be seen here, but basically it breaks down into three periods: that before 1500, that from 1500 to 1800, and that after 1800.

From 1 AD to 1500, technology does grow, but only at a very slow rate of 17.5% over 450 billion person-years (an average world population of about 300 million people times 1500 years), or an average of 0.035 percent per billion person-years. Between 1500 and 1800, the pace picks up substantially, with GDP per capita increasing by 58% in 200 billion person-years,

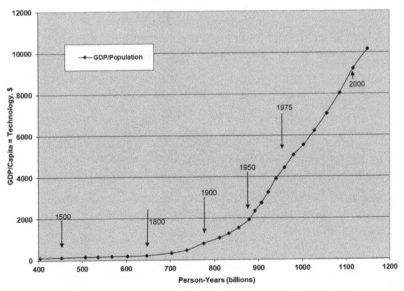

**Growth of Technology vs Global Person-Years**
**1400-2010**

**Fig. 3.3** Growth of Technology with respect to person-years, 1400-2010 (2020 dollars)

or 0.23 percent per billion person-years, a more than six-fold increase over the preceding period. Then, around 1800, technology literally takes off, with GDP per capita growing 116% over the next 90 billion person-years. As shown in Fig. 3.3, this growth continues, showing a 4700% increase over the entire 500 billion person-year span from 1800 to 2010, for an average growth rate of 0.8 percent per billion person-years.

These results make perfect sense. Before 1500 there really wasn't a world economy in any substantial sense because long-distance trade and communication were so limited. Rather than a world economy, what existed was a number of disparate civilizations including European Christendom, the Islamic world, India, China, Mexico, and Peru, each with their own economy. Important innovations made in one civilization could take centuries or even millennia to propagate to the others. For example, it took hundreds of years for such important Chinese inventions as paper, printing, and gunpowder to reach Europe, and thousands of years for European domesticated horses, wheeled vehicles, steel tools, and numerous other technologies to reach the Americas. The relevant inventive population size driving the advance of each civilization was not the whole world population, as small as it was, but the much smaller population of the civilization itself.

But around 1500, following the voyages of Columbus, Vasco de Gama, and Magellan, European long-distance sailing ships unify the world economy, creating vastly expanded markets for commerce, and making it possible for inventions made anywhere to be rapidly implemented everywhere. The effective inventive population supporting the advance of each civilization was radically expanded virtually overnight to encompass that of the entire world, creating a six-fold increase in the rate of progress per person-year compared to that of prior history. With more people engaged, the world advanced faster. Furthermore, it was precisely those countries with the greatest contact with the largest number of people worldwide, i.e., the European seafaring nations, which advanced the fastest.

Then, around 1800, the industrial revolution begins, and the average rate of progress per person-year of human effort quadruples yet again. This occurs not only because the application of steam allowed human beings to wield vastly greater mechanical power than had ever been possible before, but because particular technologies, most notably steamships, railroads, and telegraphs, radically increased the speed and thus the effective range of transportation, commerce, and communication. By the mid-1800s, innovations made by anyone, anywhere, could spread around the world virtually instantly, defining a new global reality of accelerated progress that continues to the present day. The fact that any technological advance can now have immediate global impact makes human creativity today far more powerful, and thus valuable, than ever before.[6]

The critical thing to understand here is that *technological advances are cumulative.* We are immeasurably better off today not only because of all the other people who are alive now, but because of all of those who lived and contributed in the past. If the world population had been smaller in the past than it actually was, we'd be much worse off now. Just consider what the world today would be like if the global population had been half as great in the nineteenth century. Thomas Edison and Louis Pasteur were approximate contemporaries. Edison invented the electric light, central power generation, recorded sound, and motion pictures. Pasteur pioneered the germ theory of disease that stands at the core of modern medicine. Which of these two would you prefer not to have existed? Go ahead, choose.

Human beings, on average, are creators, not destroyers. Each human life, on average, contributes towards improving the conditions of human life. This must be so, or our species would long ago have disappeared. We live as well as we do today, because so many people lived in the past and made innumerable contributions, big and small, towards building the global civilization that we enjoy. If there had been fewer of them, we would be poorer today. If we accept Malthusian logic, and act to reduce the world's population, we will not only commit a crime against the present but impoverish the future by denying it the contributions the missing people could have made.

The bottom line is this: progress comes from people. To make more progress, the world needs more people.

## PEOPLE CREATE RESOURCES

To support more people, we will need more resources. This, however, is a false problem.

Our resources are growing, not shrinking, because resources are defined by human creativity. In fact, there are no such things as "natural resources." There are only natural raw materials. It is human ingenuity that turns natural raw materials into resources. It is we who are resource-ful, not the Earth. As a result, the popular conceit that resources are perforce finite is simply untrue.

Humanity owes its power to its virtuosity in making tools. Therefore, let us consider metal, the premier material for tools' manufacture, and thus for most of history held, alongside land, as one of the two primary natural resources worth going to war for. In Table 3.1 we list the natural abundances of civilization's most important metals, in parts per million of the Earth's crust, along with the dates that each of them came into widespread use.

**TABLE 3.1 THE ABUNDANCE AND FIRST UTILIZATION OF METALS[7]**

| Metal | Abundance (ppm) | First General Use |
|---|---|---|
| Copper | 60 | 6000 BC |
| Silver | 0.075 | 5000 BC |
| Gold | 0.004 | 4000 BC |
| Tin | 2.3 | 3000 BC |
| Lead | 14 | 2000 BC |
| Iron | 56,300 | 1200 BC |
| Aluminum | 82,300 | 1890 AD |
| Titanium | 5,650 | 1960 AD |
| Silicon | 282,200 | 1960 AD |

The information in Table 3.1 should put to rest the idea that it is nature that provides our resources. For five millennia stretching from the dawn of civilization until the beginning of the first millennia BC, human use of metals was limited to a list collectively comprising less than 80 parts per million of the Earth's crust. This was because of the limitations of human technology, in particular the temperatures attainable by early kilns. But when people developed the know-how for smelting iron, our metal resource base was expanded nearly a thousand-fold. This development transformed the human prospect by making metals cheap enough to be used in common tools, instead of being restricted to aristocratic weapons and ornamental artwork. Among the tools eagerly put to use by the common folk were iron-tipped ploughs. These redefined what land could be farmed, radically expanding agriculture, population, and urbanization, which in turn greatly accelerated the rate of innovation of society overall. Then, with the development of electricity and scientific chemistry in the 19th and 20th centuries, *new metals,* including aluminum, titanium, and silicon became available. They may fairly be called new metals, because even though they were always here, and present everywhere in abundances collectively exceeding even iron many times over, they were unknown. For thousands of years migrating tribes, Roman legions, and wagon trains of westward-yearning pioneers had trod on trillions of tons of the stuff, and not one person among those myriads ever knew these immense resources existed. *Because they didn't.* They did not exist until the knowledge to create them was developed. They are creations, whose genesis is human ingenuity – resources generated by human resourcefulness.

Aluminum foil and aluminum cans have replaced tin foil and tin cans, but the significance of these new materials goes far beyond such beneficial substitutions. The new metals have enabled a new technological world, virtually magical in its capabilities compared to the culture that preceded it, featuring air travel, spacecraft, computers, global telecommunications, pocket libraries, the internet, and perhaps soon, vast amounts of inexhaustible nuclear energy.

Very well, one might admit that metals and tools are artifacts. They are indeed products of human effort, skill, and knowledge, not part of Nature's bounty. But certainly, *land* is a natural resource, right?

Wrong. Land is not a natural resource in any meaningful sense. Nature does not offer any land for our use. No fertile places exist in nature that are not already spoken for and occupied by species which are prepared to not only defend them to their deaths, but to immediately and vigorously counterattack and seize them back given the briefest opportunity. Farmland is a human creation.

Let us consider the territories currently occupied by the leading nations of Europe – France, Britain, and Germany – whose real estate prices are among the highest in the world today. In their prime, the Romans had military forces far superior to those that could be fielded by the ancient native inhabitants of these lands. They did, in fact, conquer what is now France and most of Britain, but, despite their lust for land, they did not settle those regions in any substantial way. As for Germany and the vast steppes to its east, they did not bother with them at all. Why not?

The answer lies in the Romans' (and the contemporary Celts' and Germans') technological poverty. Europe north of the Alps was of little value to the Romans, not because it was infertile, but because it was too fertile. The Roman scratch plough was fine for farming the light dry soil of Mediterranean lands, but completely inadequate for dealing with the heavy moist soils of northern Europe. Before northern Europe could be opened to large-scale civilized settlement, its farmland needed to be created. This required technologies that the Romans lacked, but which were developed during the so-called "Dark Ages" following the Empire's fall. The most critical of these inventions were the heavy plow, equipped with coulter and moldboard, and the padded horse collar. The former created a tool that could match the challenge of heavy soil, while the latter – which transferred the load on a pulling horse from its neck to its chest – made it possible to replace the Roman farmer's oxen with a much quicker and more capable draft animal. Northern European advances in iron and steel production technology were also key, as they made the axes necessary for clearing the land tougher, sharper, and far more plentiful. They also enabled the introduction and mass production of horseshoes, which greatly increased the capability of horses, both for plowing and faster land transport. The invention of the stirrup made shock attacks by mounted armored knights decisive in warfare, a development that drove the breeding of ever more powerful horses. Both horsepower and human abilities were further amplified by the introduction of the three-field system of agriculture. This was a great improvement over the Roman's two-field system, and not only because it allowed for a larger fraction of the land to be used (only one-third fallow instead of one-half). Rather, because by planting cereals in the fall and legumes in the spring (instead of just grains alternating with fallow fields) the land was made more fertile and the diet it provided more nutritious for both horses, who now got to eat oats in place of just grass, and people, whose daily fare became far richer in vegetable proteins.[8,9,10]

It was these inventions – the heavy plow, the padded horse collar, the stirrup, cheap steel, horseshoes, horse carts, stronger horses, three-field farming, and new crops, introduced between 500 and 1000 AD – that transformed northern Europe from a worthless wilderness into the heartland of Western civilization. The point requires emphasis. The terrain was always there in all its vastness and natural glory, and there had been heavily populated land-hungry societies nearby for thousands of years. Yet there was no land there worth stealing until human ingenuity created it.

Any tally of a nation's "natural resources" today would certainly include its oil and gas reserves. But this was not always so. No 18th century grand strategist would ever have given them a moment's thought. Petroleum was not originally a resource. It was always present where it can now be found, but it was not useful. For most of human history the vast majority of petroleum was undetectable and unobtainable, and that small portion that did occasionally come into view by seeping to the surface was generally considered little more than a foul-smelling nuisance that ruined good cropland or pasture. It was only in 1859 when Colonel Drake drilled the first well in Pennsylvania that oil became a resource. Marketed as "rock oil," a cheaper substitute for whale oil, "petroleum" made nighttime lighting at home much more affordable (in inflation-adjusted prices, whale oil sold in 1860 for about $150/gallon), saving several species of whales from extinction in the process. Subsequently, with the invention of internal combustion and diesel engines a few decades later, oil became the fuel for a radically expanded global transportation system, propelling vast numbers of cars, trucks, farm machines, trains, ships, and aircraft. It also became the foundation for gigantic industries producing plastics and synthetic fabrics, making clothes, and shoes, containers, and myriads of other necessities of daily life far cheaper and more plentiful than had ever been conceivable before. As if that were not enough, drilling's associated product, "natural" gas, provided first city lighting and then clean fuel for home cooking and heating, centrally generated electric power, and for any number of industrial processes as well, improving public health while saving much of the world from deforestation.

All this from a resource *which did not exist* 200 years ago!

This view of the nature of resources – as human creations – is in direct opposition to those which hold that the environment contains a finite amount of natural resources whose exhaustion entails our doom, or alternatively, that the Earth has a finite "carrying capacity" sufficient to provide the resources to sustain a limited number of people, but no more. Those views are extremely dangerous, because they portray humanity as fundamentally a race of parasites, at war with each other for rights to places in a shrinking lifeboat. Such views may seem plausible, but as the above examples show, they are entirely false. The point is critical, so let's really bang it home.

Consider the territory currently occupied by the United States of America. In the year 1500 it had a population of about 3 million people, possessing an advanced Neolithic culture. As a result of their technology, including complex languages, leather clothing, housing, efficient control of fire, weapons that could kill at a distance, fishing gear, stone tools, canoes, pottery, domesticated dogs, limited agriculture, and extensive herbal, wildlife, and weather lore, those people were able to sustain themselves in a sometimes brutally cold environment that would have been uninhabitable to their distant ancestors in the Kenyan Rift Valley. That said, they possessed almost no farmland, no metals except for insignificant amounts of natural copper, very limited textiles, no glass, and zero fossil fuel resources. Now consider the same region today, with its 330 million inhabitants, vast super-productive farmlands, and unlimited resources of iron, steel, aluminum, glass, plastics, fabrics, coal, waterpower, wind power, oil, and gas, readily capable of not just feeding and supplying itself with all necessities, but billions more besides.

It's the same place, but where it once had no farmland to speak of, now it has hundreds of millions of acres of the very best. That land was created by the pioneers and those who followed them, who cleared it, drained it, irrigated it, and did whatever else it took, including building roads and railroads to it and through it to make it useful. It once had no iron, steel, oil, or gas resources. Those have been created too.

So, what is it that "carries" us? It is not the Earth. It is human ingenuity.

## PEOPLE TIMES POWER EQUALS WEALTH

Finally, take a look at Fig. 3.4.

In Fig 3.4, I've graphed global GDP and the product of world population times total carbon use from 1920 to 2020. The result is remarkable, as over a GDP range from 1 trillion in 1920 to 80 trillion today, the two quantities track nearly perfectly. That is, if P is population, C is carbon use, and G is GDP, we have:

$$G = PC \qquad \text{GDP = Population times Carbon use} \qquad (4)$$

Why should this be? Now, one observation should be that, as much as many things have changed, we still live in the same kind of world as people did in 1920. That is, in 1920, people in advanced countries had telephones, electric lighting, mass media, and automobiles, and the bulk of energy came from fossil fuels. So, as an approximation, from 1920 to today, carbon utilization is a good stand in for energy use. Equation (4)

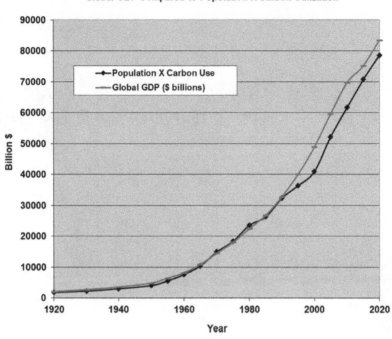

**Global GDP Compared to Population X Carbon Utilization**

**Fig. 3.4** Global GDP compared to Product of Population and Carbon Use

is saying that the rate of wealth production is proportional to the number of people engaged times the total amount of power they wield. Its significance may perhaps be clearer if it is written in the mathematically equivalent form;

G/P = (C/P)(P)    GDP/capita = (Energy/capita) times (Population)    (5)

G/P is the product per person, or standard of living. C/P is energy per capita, a measure of the energy richness of a society. So, if we want to substantially increase living standards, we will need to have more people, with more power available for each person to use.

This can only be accomplished using nuclear energy.

# CHAPTER 4

# WHAT IS NUCLEAR ENERGY

$$E=mc^2$$
**– Albert Einstein**

**IN THE LATE** 1800s, scientists faced an apparently insolvable problem. According to all known laws of science, the Earth was impossible. Based on studies of erosion, sedimentary features, and ever more convincing fossil evidence for a long history of prehistoric evolution, geologists and biologists, notably including Charles Darwin, had come to the conclusion that the Earth had to be at least 300 million years old. But physicists, who could measure the energy output and mass of the Sun, said that could not be, because even if the Sun were made of solid coal, burning at its current rate, the longest it could possibly have lasted would have been about 3000 years. The great scientist, William Thompson (Lord Kelvin) tried to find a way around that by postulating that the Sun was not powered by chemistry, but by infalling meteors and gravitational collapse, but the longest lifetime that theory could provide was about 30 million years. Darwin's theory of evolution, he therefore declared, could not be valid. The Earth simply hadn't existed long enough.[1]

Except it had, and the evidence kept piling up to prove it. In fact, Darwin and the geologists had actually greatly *under-estimated* the age of the Earth, and hence the Sun, which is close to 4.5 billion years old, and there are other stars that are over 10 billion years old. Had these facts been known at the time, they would only have made the existence of the Earth, the Sun, and the rest of the universe even more obviously impossible.

But the Earth, the Sun, and the universe are not impossible, because the stars are not powered by coal, gravitational collapse, or any other form of energy familiar to classical physics. They are powered by nuclear forces, which are *millions of times* larger in magnitude then the electrical forces that provide the energy source for chemical reactions.

Hints of the potential existence of such more powerful forces of nature came in 1896 when Marie and Pierre Curie discovered radioactivity. The New Zealand born physicist Ernst Rutherford then measured the energy of alpha particles emitted by radium to be an astonishing several million volts, or a million times the several-volt energy of typical chemical reactions. Commenting on this finding in 1904, he said:

"The discovery of the radio-active elements, which in their disintegration liberate enormous amounts of energy, thus increases the possible limit of the duration of life on this planet, and allows the time claimed by the geologist and biologist for the process of evolution."

Then in 1905, Albert Einstein presented his theory of Special Relativity, one of whose findings is that matter and energy are ultimately two different forms of the same stuff, with the conversion rate between the two given by his famous formula $E=mc^2$, or energy equals mass times the speed of light squared. Now the speed of light is 300 million meters per second, so according to Einstein, 1 kilogram of mass – of any kind, gold, dirt, water, or rotten tomatoes – is the energy equivalent of 90 billion Megajoules (MJ), or 25 billion kilowatt hours. That's enough energy to power the entire world for an hour.

Well, that is what we would get if we could convert the entire kilogram from matter to energy. We are not quite clear on how to do that yet, but a way to get a small (but still huge!) fraction of that enormous amount of energy was made clear in 1920 by British physicist F. W. Aston, who made careful measurements of the mass of the nuclei of many elements. Nuclei consist primarily of positively charged protons and neutrally charged neutrons, each with one atomic mass unit. The number of protons is the atomic number, and as it is matched by an equal number of electrons orbiting the nucleus, it determines the chemical identity of that atom. A given chemical element always has the same number of protons, but the number of neutrons in its nucleus can vary, with each different number of neutrons defining a different isotope of that element. For example, common hydrogen just has one proton in its nucleus, but there is an oddball minority isotope (it's about 1 out of every 6000 hydrogens on Earth) called deuterium that has both a proton and a neutron. Added together, the combined number of protons and neutrons determine the atomic weight of that isotope.

Well almost, but not exactly. The atomic weight of deuterium is not 2.0000, but 2.0141, and the atomic weight of common helium, with 2 protons and 2 neutrons, is not 4.0000 but 4.0026. Two plus two is supposed to equal four, but if you combined two deuteriums to make a helium, instead of getting a nucleus with a weight of 4.0282 the result only weighs 4.0026. You've lost 0.0256 units of mass – 0.64% of the 4-unit total! That's possible, because the four nucleons together pack more efficiently than they do as two pairs. So, the extra packing material, which is manifested in each pair as mass, can be released as energy when they are grouped as a foursome. The important thing is that Aston found that this is an example of something that holds true much more broadly. Compared to the mass per nucleon of the lightweight atoms like hydrogen or the heavyweight atoms like uranium, there is *mass missing* from the nucleus of the middle-weight elements, like iron, which has atomic mass 56. The missing mass, Aston reasoned, is turned into energy and released when such nuclei are assembled.

Imagine a cruise ship which offers maximum discounts to groups of 56, but charges more per ticket the further a group booked together is away from that ideal size. In the fusion example, cited above, two couples saved money by booking together as a foursome. That saved them cash, i.e., energy, so they could spend more on drinks during the trip. A group of 224 could also save money by splitting into four groups of 56, or four unequal medium-sized groups averaging out to 56. So sure, split up the group, get some extra cash and party! But once you are in a group of around 56 you will need to pay extra to either split it up into smaller parties or join together with another to form a larger group. So unless you can get cash from the outside world, you are stuck, because you've already used up your potential 56-person group discount. Similarly with nuclei, the more mass lost, the more tightly bound the resulting nuclei would be. This concept led to what came to be called "the curve of binding energy," a modern version of which is presented in Fig. 4.1.

Iron (Fe, atomic number 26, atomic weight 56) is at the top of the curve, because it has the largest energy deficit, or "binding energy." But really the curve is drawn upside down. It should be the other way around, with iron at the bottom of an energy hole, with a sharp slope going down to it from hydrogen on the left-hand side, and a gentle slope leading down towards it from uranium on the right-hand side. If you were to then release a ball from either end of the curve, it would roll down towards iron, speeding up and releasing energy as it went, until it reached the bottom and got stuck.

That's basically how atomic energy works. As soon as Aston published, the great British physicist Sir Arthur Eddington

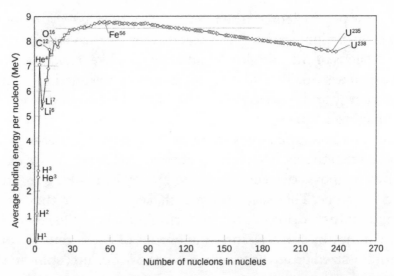

**Fig. 4.1.** The Curve of Binding Energy. From an energy point of view, it is really upside down, with the middle weight elements like iron (Fe) at the bottom of an energy hole. Energy can be released by rolling down into the hole from either extreme. (Credit: Wikipedia Commons)

realized that he had found the mechanism that powers the Sun: hydrogen fuses to form helium, releasing binding energy in the form of heat and light. Later, Russian-born physicist George Gamow was able to elaborate this insight into a comprehensive theory of stellar evolution, with young suns burning hydrogen into helium, and then helium into oxygen and carbon, and ultimately combining these to make iron in their old age.[2]

But you can also release nuclear energy by rolling into the middleweight pit from the other side, by splitting heavy nuclei like uranium into smaller fragments. So, there are two ways to go: nuclear fusion or nuclear fission. As the curve shows, you can get more energy out of fusion, which converts 0.4% of the total reaction mass into energy. This compares to about 0.1% energy yield from fission fuel. These may be small fractions to the total energy that is there, and, from a practical point of view the useful fraction is even less, because the isotopes of various elements that we know how to make use of as fusion or fission fuels are typically small minorities of the more common kinds of their chemical element. But it hardly matters. At the end of the day, the energy yield from either of these processes is vastly more than anything possible with chemical fuels.

Take a kilogram – about a quart – of water for example. It consists of 111 grams of hydrogen and 889 grams of oxygen. If you had that much hydrogen and oxygen, mixed them together, and threw in a match, they would light explosively, releasing 15.4 MJ of energy, about as much as you would get burning 1/10th of a gallon of gasoline. But of those 111 grams of hydrogen, about 0.037 grams would be deuterium, an unusual type of hydrogen that in addition to the single proton in its nucleus contained by normal hydrogen, also has a neutron. If you burned that 0.037-gram speck of deuterium in a nuclear fusion reactor, you would release 13,700 MJ – 900 times are much as you got by combustion of the whole kilogram of hydrogen and oxygen, or about as much as you would get by burning 90 gallons of gasoline!

Think about that. If you had a fusion-powered car that got 30 miles to the gasoline-gallon of energy, you could travel 10,800 miles on the fusion power available from the tiny amount of deuterium present in any gallon of water!

So, energy is released either when light nuclei like hydrogen or deuterium (atomic weights 1 and 2) combine to make heavier nuclei, or when heavy nuclei like uranium or plutonium (atomic weight 238 and 239) split to make lighter nuclei. The former process is known as nuclear fusion. It is what powers the Sun and all the stars. The latter process is known as nuclear fission. Fusion energy can be released using very common elements for fuel. But it is hard to do because atomic nuclei, being all positively charged, repel each other, so it takes a lot of energy to force them to come close enough to each other to react. Fission reactions, on the other hand, only need to split big nuclei that have already been put together for us by nature (using the power of ancient supernova star explosions). Fission does require fuels that are harder to come by than those that drive fusion, but it is much easier to do, and so is the kind of nuclear energy that can be used to produce commercial electricity now.

The most important thing there is to know about nuclear power is that it is far and away the greatest energy resource available to humanity today, exceeding all others combined thousands of times over. Now energy is the most basic physical resource. With sufficient energy, you can make anything. In all chemical reactions, such as burning fuel or rusting metal, no matter is ever used up. It is simply rearranged. Despite all the food, fuel, and metals that people have used through all of history, the universe is not one kilogram shorter in any of the elements that compose these materials. When we say something has been consumed, what we really are saying is that its elements have been rearranged from a useful form into a less useful form. With enough energy, you can put them right back. In arguing for limits to growth, Malthusians always inevitably end up pointing to energy. There is only so much to go around, they say, so human aspirations must be crushed. Lest we run out of energy, they claim, people in advanced countries must accept lower living standards and the poor nations must stay poor forever.

But nuclear power completely upends any such rationale for putting chains on humanity. This is shown in Table 4.1, which presents both the size and per capita cash value of the Earth's energy resources. Table 4.1 is divided into two sections, the first dealing with resources that can be used today, the second showing additional resources that we can create in the future provided we continue to advance our technology. It can be seen that nuclear resources vastly outclass fossil energy sources in both the current and future categories.

**TABLE 4.1 ENERGY RESOURCES WE CAN CREATE ON EARTH[3]**

| | Energy (TW-years) | Value per Capita |
|---|---|---|
| Current Resources | | |
| Known Conventional Fossil Fuels | 1500 | $114,000 |
| Estimated Unknown Conventional Fossil Fuels | 3000 | $228,000 |
| Nuclear Fission (Uranium ore, without reprocessing) | 700 | $53,200 |
| Nuclear Fission (Uranium ore, with reprocessing) | 100,000 | $7,600,000 |
| Nuclear Fission (Thorium ore, with reprocessing) | 400,000 | $30,400,000 |
| Future Resources | | |
| Natural Gas (including sub-sea methane hydrates) | 15,000 | $1,140,000 |
| Nuclear Fission (Uranium from sea water, no reprocessing) | 24,000 | $1,825,000 |
| Nuclear Fission (Uranium from sea water, with reprocessing) | 3,430,000 | $260,085,000 |
| Nuclear Fission (thorium from sea water, with reprocessing) | 13,720,000 | $1,043,406,000 |
| Nuclear Fusion | 400,000,000,000 | $31,200,000,000,000 |

Focusing on the resources that are usable today, we see that our known fossil fuel reserves are sufficient to provide 4,500 terawatt-years (TW-years) of electricity. (A watt-year is an amount of energy, and a terawatt-year is a trillion watt-years. For example, a 100-watt light bulb burned for a year uses 100 watt-years of energy.) Since humanity is currently using about 20 TW-years annually, fossil fuel reserves would represent enough energy to power the globe at current rates for 225 years, or 45 years at five times our current consumption. (This figure is meant only to provide scale, and implies no prediction of a time to fuel exhaustion, as usage rates will change and new fuel reserves are always being discovered. But still, they show why fossil fuels are sufficient to power today's world, but are not abundant enough to power a grand flourishing human future.) By contrast, our presently known reserves of uranium and thorium ore comprise at least 500,000 TW-years of energy, over a hundred times as much as fossil fuels, sufficient to power the globe at ten times current rates for over 2,500 years. These could be multiplied many times over by employing the uranium and thorium found in granite or seawater. Indeed, if burned in a fission reactor, the 2 ppm of uranium and 8 ppm of thorium within in a typical block of ordinary granite would yield a *hundred times* as much energy as that obtained by burning a mass of oil equal to that of the rock.

Using fusion fuels, however, we could go on for millions of years at thousands of times current levels – and that's only using the deuterium available on *this* planet. There are other planets – and other potential fusion fuels – that could multiply this already enormous figure beyond any meaningful measure.

We are literally surrounded by mountains and oceans of energy.

There are those who argue that the Earth's natural resources should be considered "the common heritage of mankind." This is a debatable proposition, but it has some merit, since they are in fact a natural gift of great value to everyone. Furthermore, regardless of current ownership, you should care about how much there is, because your money is only valuable if there is stuff for you to buy. (This is why you are richer than John D. Rockefeller. You can buy things that he couldn't.) So, if we were to divide our planet's energy largesse equally, how much would your share of it be worth? The answer is given in the second column of Table 4.1, showing the cash value of these resources. These figures come from taking the total worth of each resource if converted into electricity at a price of 7 cents per kilowatt-hour, and dividing it among the globe's current population of around 8 billion people. Figured this way, the share of the world's common heritage of fossil fuels fairly due to you and your all heirs would be worth about $340,000. That may sound like a lot, but if you and your folks want to be able to use, say $3,400 worth of energy per year, it would only cover you for a century. In contrast, the value of the energy your nuclear reactor fuel could provide would come to some $38 million.

Of course, you and your grandkids only get your $38 million worth of energy if we *have* nuclear reactors. Without them, all of your uranium and thorium is just so much dirt. I put it this way because it is important that people understand exactly what they are giving up should we be forced to forgo nuclear power.

# CHAPTER 5
# HOW TO MAKE NUCLEAR ENERGY

**IN ORDER TO** understand how to make nuclear energy, there is no better way than to learn how it was figured out in the first place. Let's start at the beginning.

In 1898, Marie and Pierre Curie discovered the element radium. Radium, symbol Ra, has many isotopes, but the most common one is radium-226 (also written $^{226}$Ra), so-called because it has 88 protons — as all radium does, and 138 neutrons, adding up to 226 in all. The Curies found that radium-226 decays to radon-222 with a half-life of 1600 years, emitting an alpha particle containing 2 neutrons and 2 protons — essentially a helium nucleus — with an energy 4.8 million volts of energy in the process. That's huge. By comparison the energy released when you react hydrogen with oxygen is about 1.2 volts. So you can imagine how amazed they — and subsequently the world at large — were. The Curies had discovered a type of reaction releasing energy millions of times greater than the most energetic reactions known to chemistry. Furthermore, subsequent investigations showed that the radon-222 then decayed with a half-life of 3.8 days to polonium-218 (which Marie named for her homeland), which then decayed with half-life of 3 minutes to lead, with further multi-megavolt (MeV) alpha particles released at each step.

That was pretty exciting. But while the energy released per reaction was enormous, the number of reactions involved in such natural radioactivity was so low that the overall rate of energy release was too small to be of much practical value. Radium decays with a half-life of 1600 years, which means that it takes sixteen centuries for a sample of radium to release half its energy. So, while a gram of radium might contain as much energy as five tons of coal, spread over 1600 years it releases as much heat as you would get by burning 360 milligrams (about the weight of an aspirin) of coal per hour. In his 1913 novel, *The World Set Free*, writer H.G. Wells dealt with this difficulty by having scientists figure out a way to speed this up a million times over. But that was science fiction. In real life, all radium seemed good for was lighting up watch dials.

High-energy alpha particles can do more than make fluorescent chemicals glow, however. They can knock neutrons loose from other nuclei, for example those of beryllium. This is the trick that James Chadwick managed to do in 1932, using polonium as his alpha source. It took a bit longer for people interested in enabling nuclear energy to realize the significance of this, because the odds of such an event are low, and it takes thousands of alphas hitting a metal target to knock loose a few neutrons. That's because alpha particles are positively charged, just like any other nucleus. So they try to swerve away as they approach their targets, making it hard to score a hit. But here's the thing: neutrons have no charge, so they are not repelled by target nuclei. In fact, they can even be attracted to them. Furthermore, when they strike home, they don't disappear, they multiply. This makes them *much* better fission bullets than alphas.

Even using neutrons, however, there is only one isotope found in nature that is easy enough to split to make it a practical fuel for nuclear fission. This is uranium-235, which constitutes 0.7% of all natural uranium (the rest is uranium-238). In addition, there are several artificial isotopes, most importantly uranium-233 and plutonium-239, that are also "fissile." Under the right conditions, each of these will undergo a very energetic fission reaction when they absorb a neutron, exploding into two medium-sized nuclei fission fragments, plus two, or occasionally three, super-hot MeV class "fast" neutrons as well. If one of those new neutrons manages to start a fission in a neighboring uranium or plutonium, still more neutrons will be born, which can generate still more if they cause fissions, thereby setting off a chain reaction, multiplying the numbers of neutrons and thus the reaction rate exponentially.

In order for this to work, you need to have a *critical mass* of the fissile material, because if you have too little, most of your neutrons will fly out of the system without scoring a hit, and you need to have your neutrons average at least one kid each in order for a chain reaction – or "critical" reaction – to be maintained. The critical mass can be minimized by having an optimum geometry for the fissile system, packing neutron-reflecting materials like graphite or beryllium around it, putting in some "moderating" material to slow the neutrons down to increase their chance to interact with a fissile nuclei before exiting, and eliminating non fissile impurities that have a hankering to gobble up neutrons before they can reproduce. But provided you can, by hook or crook, engineer a critical assembly, all you need is a little radioactive neutron source starter kit to light it up.

The story of how all this was discovered is absolutely fascinating, and is told brilliantly in Richard Rhodes magisterial book, *The Making of the Atomic Bomb*.[1] But briefly, it was the exiled Hungarian-Jewish physicist Leo Szilard who, while bumming around London in the early 1930s realized that neutrons that might be used to launch a chain reaction. Accordingly, on July 4, 1934 – the same day that Marie Curie passed on to the other world – he filed the patent for nuclear energy.

**Fig. 5.1** Lise Meitner and Otto Hahn in the lab at the Kaiser Wilhelm Institute in 1913. Meitner was the second woman in the world to earn a PhD in physics. Two years after this photo was taken, she (like Marie Curie, on the other side) was at the front, using X-rays to save the lives of wounded soldiers. (Credit: Wikipedia Commons)

However, Szilard had no idea of how to make it work. He thought, along with the rest of the physics community, that neutrons could solely be used to chip away other neutrons from target nuclei, which is not the same thing as nuclear fission. Furthermore, he was clueless as to what the right target nuclei might be (he thought beryllium.) As the 1930s wore on, he tried all sorts of experiments on beryllium, then other elements, without useful results, and became so demoralized that in late 1938 he wrote the British admiralty (with whom he had filed the patent, in order to keep it secret from the Nazis) to withdraw his patent.

But even as Szilard's letter was in the mail, nuclear fission was discovered.

Lise Meitner, an Austrian of Jewish heritage, had fled to Sweden to escape the Nazis. (The fact that her family had converted to Christianity and that she had served patriotically as an X-ray technician with the Austrian army on both the Italian and Russian fronts during the Great War did not protect her.) Shortly before Christmas 1938, she received a letter from her pre-Nazi scientific collaborator at the Kaiser Wilhelm Institute, Otto Hahn, asking her help in explaining some inexplicable results. According to Hahn, a chemist, after bombarding some uranium with neutrons he had discovered traces of barium (atomic number 56, weight 137) and lanthanum (atomic number 57, weight 139) in the sample. You'd have to knock off nearly a score of alpha particles from a uranium (atomic number 92, weight 238) nucleus to chisel it down to one of these elements. Was that possible?

Meitner took the letter with her to an out-of-town Christmas party, where she was joined for the holiday by her nephew, Otto Frisch, who was working in Copenhagen with the Danish physicist Niels Bohr. Discussing the results with Frisch, Meitner had a flash of insight. *Why did uranium have the largest nucleus of any element?* It must be because anything larger was unstable! By getting uranium nuclei to absorb neutrons, Hahn had destabilized them, causing them to split, or fission, just like an amoeba that had grown beyond its maximum allowable size.

Meitner did some calculations. If a uranium split, the two resulting large nuclear fragments, each with 30 to 60 protons, would repel each other forcefully, releasing something like 200 MeV of energy – far more than a 5 MeV alpha emission. She then added up the mass of the fission fragments that would be produced. They amounted to less than that of the original uranium nucleus by about $1/5^{th}$ of a nucleon. Converted to energy via Einstein's formula, that equals 200 MeV. Eureka.

Returning to Denmark, Frisch told Bohr about the discovery, just before the physicist set sail for New York to attend a scientific conference.

Bohr was famous for having made fundamental discoveries about the structure of the atom and quantum mechanics. Arriving in January 1939 he was greeted warmly at the dock by some the world's scientific elite, including Szilard and Italian physicist Enrico Fermi (who had fled Mussolini-ruled Italy because his wife, Laura, was Jewish). Bohr told them of the discovery. The two were astonished, and not quite convinced. Fermi resolved to put it to experimental test. His Columbia

University graduate student, Herbert Anderson, did so, as did Richard Roberts, working in the Washington DC Naval Observatory. Before the conference (which was held in DC) was even over, both had obtained positive results. Roberts invited the attendees, which included not only top scientists but reporters for the *Washington Star* and the *New York Times* to come to the observatory and see his experiment in action. There was no confusing the oscilloscope traces of the 100 MeV-class fission fragments with those of 5 MeV alpha particles. Nuclear fission was a fact.

But Szilard and Fermi realized that there was more to the discovery than just exciting new physics. As you move up the periodic table, the ratio of neutrons to protons in nuclei increases. Hydrogen, atomic number 1, usually has 1 proton and no neutrons. The most common isotopes of helium, carbon, and oxygen (atomic numbers 2, 6, and 8) have equal numbers of protons and neutrons. When you get to iron, the neutrons outnumber the protons 30 to 26, and this preponderance of neutrons to protons continues to increase as the nuclei get heavier, with the ratio in uranium being 146 neutrons to 92 protons. If you split uranium into lighter weight nuclei, there are going to be too many neutrons to fit into the major fragments. Those extra neutrons would be set free, and once free, would be able to start additional fissions. That meant one thing: a chain reaction was possible.

Szilard quickly sent a letter to the British admiralty, requesting that they please ignore his previous note withdrawing his patent for the chain reaction. Fermi then attempted to convince the US Navy brass that they needed to take an interest in this potent new energy source, but he was dismissed as "a wop." So Szilard drew up a second letter, and in order to make sure it would be read, had the most famous scientist in the world sign for him. That letter, with Albert Einstein's signature on it, Szilard then had hand-carried to President Franklin Roosevelt by influential financier Alexander Sachs, to advise him that it was now clear that the laws of physics allowed the construction of atomic bombs. Furthermore, physicists working for Hitler unquestionably knew that too.

FDR quickly got the drift. "So, what you are saying, Alex," he reportedly replied, "is that you don't want Hitler to blow us up." Sachs nodded.

Thus was born the Manhattan Project.

While the laws of physics allow anyone to make atom bombs, in order to actually do it you need to have a critical mass of fissile material. It is impossible to make a bomb out of natural uranium, because natural uranium is only 0.7% fissile uranium-235. The other 99.3% is uranium-238, which interacts with neutrons by absorbing them. The uranium-239 thus formed is unstable, and within a few days will decay by successive emission of two beta particles (which are really just electrons) into neptunium 239, then plutonium-239. (It's Uranus, Neptune, Pluto – get it?) Plutonium-239 is fissile, but making an atom of it costs you a neutron, and you can't afford to spend most of your fission product neutrons (usually two, sometimes three per fission) and still maintain a critical chain reaction.

So if you want to make an atom bomb, there are a couple of things you can do. You can separate out uranium-235 from natural uranium, to create *enriched uranium* that is much more reactive. You can make uranium metal that has been enriched to about 20% uranium-235 go critical, but it wouldn't make much of a bomb, because the rate at which the neutrons multiply would be slower than speed at which the bomb would disassemble, once it started to get hot. The explosive yield of such a crude bomb would be very low. To get a truly impressive yield, you need to enrich the fissile material to about 80 percent (as the Hiroshima bomb was), and ideally over 90 percent, as modern nuclear weapons are. Furthermore, you need to bring the critical mass together faster than the reaction can happen, but then make the neutrons multiply very fast, so most of the uranium's energy can be released before the weapon blows apart. That is a very tough trick, which I will not explain because all the people who should know how to do it already do.

Enriching uranium isn't easy, because uranium 235 and uranium-238 are the same element, and so behave identically in all chemical reactions. The only way to separate them is by their weights, which differ by less than 1.3 percent. The techniques involved, which include gaseous diffusion systems, mass spectrometers, and staged centrifuges, require huge amounts of energy. Fortunately, this was available to the Roosevelt administration from the newly built Tennessee Valley Authority, whose hydroelectric power generation capacity greatly exceeded the consumer demand of its not-yet industrialized region. Still, in order to maximize the useful power available, FDR agreed to the request of General Leslie Groves, the Manhattan Project manager, to provide silver from the US mint to make wire with the minimum possible electrical resistance.

Plutonium is easily separable from uranium, because it is a different element. As mentioned earlier, plutonium-239 is also fissile, and you can make it from common uranium-238 if you have spare neutrons. The only way to get enough of these to be useful, however, is to run a steady state critical nuclear reactor. But how can we make a reactor critical if all we have to start with is low enriched natural uranium?

Enter Enrico Fermi. In 1935, he was conducting some experiments at his university in Rome involving bombarding substances with neutrons to see if he could induce radioactivity. He was getting good results, particularly with uranium targets, when he was forced to move his experiments from the marble table he was using (this was Rome), to a wooden table. Incredibly, the number of counts registered by the Geiger counter doubled when the experiment was moved to the wooden table! The obvious suspect was experimental error, but no, the effect

**Fig. 5.2** Enrico Fermi (Credit: US Department of Energy)

was reproducible. As soon as the experiment was moved back to the marble table the counts were cut in half. Then they doubled again when it was returned to the wooden table. What could possibly cause such a bizarre effect?

Fermi figured it out. Wood is made of light atoms, including a lot of hydrogen, plus carbon and oxygen. Marble is dominated by heavier atoms, like calcium, with very little hydrogen. A neutron hitting a hydrogen nucleus – i.e. a proton – something with close to the same mass as the neutron, will share its energy with that mass almost equally, while one hitting a heavy calcium would just bounce off, like a ball hitting a brick wall. (This is because in a collision between particles, the energy of a collision is divided between the particles afterwards in inverse proportion to their mass.) Fermi reasoned that the picture of neutrons causing atomic reactions by crashing into nuclei like cannonballs smashing them to pieces was incorrect. Instead, they should really be viewed as wanderers, which can be attracted into a nucleus if they come close enough to one for long enough, so the *slower* a neutron is traveling, the more chance it has of causing a reaction.

This theory, that slowing, or "moderating" a neutron would increase its atomic reactivity was correct. Its discovery won Fermi the Nobel Prize in 1938, which proved really handy, since he used that as an excuse to travel with Laura out of fascist Italy to Stockholm to pick up his prize money, after which the two of them immediately split for New York, and freedom. But the significance of the discovery for nuclear engineering is that it made it possible to build the first critical nuclear reactor.

Fission neutrons are born with lots of energy – about 2 MeV – which means they are born fast. But if they experience collisions with other particles close to their own mass, they split their energy with them. That slows them down, increasing their reactivity. If they collide enough times, their energy will be reduced to that of the surrounding medium. By cutting their energy down by a factor of 80 million, from 2 MeV to 0.025 eV (the energy equivalent of room temperature) this "thermalization" process will reduce the neutron's speed by a factor of 9,000 (because the energy of a particle goes in proportion to its velocity squared). This will increase their reactivity *a lot*, so much so, that even natural uranium can be made to sustain a chain reaction. You can't build a bomb this way, however, because the rate at which neutrons can multiply if they have to go through dozens of moderating collisions before they react is way too slow.

If every atom of uranium in a bomb were to fission – i.e. if the bomb were to achieve 100 percent yield – you would get 200 MeV per atom. But well before the neutrons have multiplied to generate 200 MeV per atom, they will have produced 1 eV per atom, which is enough energy to raise the bomb material to over 10,000 degrees Kelvin. (1 eV equals 11,000 K). At that temperature, the uranium and everything else the bomb is made of will be very hot gases, expanding at a speed on the order of 1000 meters per second – the speed of a projectile leaving a high velocity gun. So in 0.1 milliseconds, the bomb materials will have expanded in all directions by ten centimeters, which is more than enough to make the reaction go subcritical and shut down. But it takes milliseconds for newborn fast neutrons to go through enough collisions to become slow neutrons. As a result, slow neutrons simply can't multiply fast enough to produce even a bomb yield of 1 millionth of a percent. The bomb would disassemble long before they could do so, leaving you with a maximum energy yield no better than what you could get from a vastly cheaper chemical explosive.

In fact, it's worse than that, because you can't cool a neutron down to 0.025 eV if all the materials it might touch are much hotter than that. Furthermore, even if you could, once the bomb materials are hotter than a few eV they will be flying apart faster than a slow neutron can travel. So, let's say you had a team of top-notch magicians, including Merlin, Gandalf, and Hermione Granger to help you, and they used magic to make a zillion cold neutrons appear in the middle of super-hot expanding gas. It still wouldn't do you any good, because the cold neutrons wouldn't be able to catch up with the fleeing hot uranium nuclei in order to split them.

You just can't make a bomb using slow neutrons. But you can build a reactor that way, and that is exactly what Fermi did.

Viewed in the light of today's knowledge, Fermi's nuclear reactor was a very simple device, and many readers might want to build one of their own. This is entirely possible. It's not rocket science after all. But before you do, you will need to learn a little more about nuclear physics. (You should also get permission, in writing, from your state and local governments, as they may have zoning laws or other regulations that forbid the operation of critical nuclear reactors in your neighborhood. If you are a minor, be sure to get permission from your parents as well.)

But back to the physics. The question of whether a neutron will cause a fission when it travels through a material containing a lot of target nuclei, or do something else, for example scatter elsewhere or be absorbed without fission, is one of probability. To understand this, imagine that you are a kid who wants to launch model rockets. You live in a growing suburb, but there are still a few farms within bicycle range, so you choose one of them for your launch site. What is the chance that when your rocket comes down it will hit the barn?

Well, the barn is pretty big, so there is a real possibility that you might hit it. Let's call that probability 1 barn. The house is smaller, so maybe the probability of hitting it is half a barn. The vegetable garden is four barns, the horse corral, ten barns, the farmer's car, 0.01 barns.

In the same way, nuclear physicists also measure the probability of scoring a certain kind of hit on a nucleus in units of "barns." Except that while your kid rocketeer's barn units might be equivalent to 300 square meters of cross-sectional target area, nuclear physicist's barns are somewhat smaller, with 1 barn equaling one trillionth of a trillionth of a square centimeter (1 barn $=10^{-24}$ cm$^2$.)

What is the cross section, in barns, of various nuclei? There are three different nuclear reactions of interest: fission (which releases energy and produces neutrons,) absorption (or capture, which kills neutrons) and scattering (which thermalizes neutrons). You can see the value of these, in barns, for the interaction of slow and fast neutrons with the most common materials used in nuclear engineering in Table 5.1.

## TABLE 5.1 CROSS SECTIONS OF SELECTED NUCLEAR ENGINEERING MATERIALS (BARNS)[2]

| Material | Thermal Cross Section (barns) | | | Fast Cross Section (barns) | | |
|---|---|---|---|---|---|---|
| | Scattering | Absorption | Fission | Scattering | Absorption | Fission |
| Moderators | | | | | | |
| Hydrogen-1 | 20 | 0.3 | 0 | 4 | 0.00004 | 0 |
| Hydrogen-2 | 4 | 0.0004 | 0 | 3 | 0.000007 | 0 |
| Carbon-12 | 5 | 0.002 | 0 | 2 | 0.00001 | 0 |
| Oxygen-16 | 4 | 0.0001 | 0 | 3 | 0.00000003 | 0 |
| Structure | | | | | | |
| Chromium-52 | 3 | 0.5 | 0 | 3 | 0.002 | 0 |
| Iron-56 | 10 | 2 | 0 | 20 | 0.003 | 0 |
| Nickel-58 | 20 | 3 | 0 | 3 | 0.008 | 0 |
| Zirconium | 8 | 0.18 | 0 | 5 | 0.004 | 0 |
| Absorbers | | | | | | |
| Boron-10 | 2 | 200 | 0 | 2 | 0.04 | 0 |
| Cadmium-113 | 100 | 30,000 | 0 | 4 | 0.05 | 0 |
| Xenon-135 | 380,000 | 2,714,000 | 0 | 5 | 0.0008 | 0 |
| Fuels | | | | | | |
| Thorium-232 | 13 | 7.6 | 0.0002 | 5 | 0.09 | 0.07 |
| Uranium-233 | 12 | 45 | 531 | 3 | 0.01 | 2 |
| Uranium-235 | 10 | 99 | 583 | 4 | 0.09 | 1.2 |
| Uranium-238 | 9 | 2.5 | 0.00002 | 5 | 0.07 | 0.2 |
| Plutonium-239 | 8 | 269 | 748 | 5 | 0.05 | 2 |
| Plutonium-240 | 10 | 286 | 0.03 | 5 | 0.09 | 1.4 |

Looking at Table 5.1, you can see that the only fuels that are any good are uranium 233 and 235, and plutonium-239. Of these, only uranium-235 exists in nature, as 0.7% of natural uranium. As you can see, the fission cross section of uranium-235 increases 500 times over if you moderate your neutrons from fast to slow. If you had very heavily enriched (over 80%) fuel, the U-235 fast fission cross section of 1.2 barns would be just good enough to achieve criticality. But please don't do that at home. Highly enriched U-235 is ridiculously expensive, and if you are caught with a critical mass, the judge might not buy your Second Amendment defense. It's better to go with natural uranium, just as Fermi did. But with the 0.7% enrichment of natural uranium, you are going to need to moderate your neutrons. How should you do it?

When particles collide, they split the collision energy in inverse proportion to their mass. So if lightweight particles collide with heavy ones, the lightweights bounce off with most of the energy. But when lightweights collide with each other, they split their combined energy evenly. Therefore, if you want to moderate neutrons, which have an atomic weight of one, you should aim to use the lightest atomic weight moderator materials you can. That would make hydrogen-1 the obvious choice. But hydrogen-1 – the overwhelming majority of the hydrogen found in normal water – also poses a problem; it has a certain proclivity – 0.3 barns – to absorb neutrons, and a natural uranium-fueled reactor has no neutrons to spare. Every 583 barns worth of fission cross section offered by a uranium-235 goes along with another 99 barns of absorption cross section in the uranium-235, plus another 357 barns from the uranium-238 (because there are 143 uranium 238 nuclei in natural uranium for every U-235). That adds up to 456 barns of absorption in the fuel itself, almost as much as there is for fission. If the volume of your water moderator equals a few times the volume of your uranium, you are likely to have several hundred times as many hydrogen atoms as U-235's, and at 0.3 barns each that could add up to more than another hundred barns worth of absorption. If you put that together with the 456 barns worth of absorption in the fuel, plus a few more in the reactor structure, the absorption cross section of your reactor material will be greater than its fission cross section. That means that the majority of your neutrons will be killed by absorbers before they can ever reproduce, and if you put those losses together with those that will happen because some neutrons will wander outside the reactor through its edges, the bottom line is that you will never achieve criticality, and you will lose your science fair prize.[3]

Well, Fermi could not afford to lose his science fair prize, so he looked around for a better moderator than water. He found it in the form of graphite.

With atomic weight 12, graphite takes six times as many collisions as hydrogen to thermalize a neutron (because hydrogen takes away half of the collision energy, whereas a carbon takes away a twelfth), but its 0.002 barn absorption cross section is less than 1 percent as great. Fermi calculated it could do the job.

Under Manhattan Project sponsorship, Fermi pulled together a crack team of some of the top talent in American physics to create the first nuclear reactor. One team member, Samuel K. Allison, a professor at the University of Chicago, thought it would be a good idea to put it under the stands of the university's Stagg Field football stadium. So that's where it went.

(Allison, a protégé of Arthur Compton, subsequently became the graduate school advisor of Fred Ribe, who later became the head of the controlled thermonuclear fusion program at Los Alamos. Ribe was my grad school advisor. So, in a sense, Allison was my academic grandfather. I share his dislike for spectator sports. Of my two biological grandfathers, one was a wit and the other was a hussar. I may have gotten something from them too.)

The reactor, dubbed Chicago Pile-1, was assembled in November 1942. It contained 4.9 metric tons of natural uranium metal and 41 tons of uranium oxide for use as fuel, and 330 tons of ultra-pure graphite for moderator. To keep the reactor from going critical on its own, rods made of cadmium, which has an enormous 30,000 barn neutron absorption cross

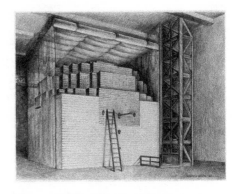

**Fig. 5.3** Chicago Pile-1 Under Construction (Credit: Argonne National Laboratory)

section, were put in it. To make it go, all you had to do was pull the rods out.

On December 2, 1942, Fermi gave the order to pull the rods. The experiment began in the morning but was aborted when an

**Fig. 5.4** On December 2, 1942, the rods were pulled out and Chicago Pile-1 went critical. (Credit: US Atomic Energy Commission)

over-sensitive safety system caused a control rod to be dropped into the reactor.

Then, that afternoon, with Allison standing by with a bucket of concentrated cadmium nitrate to throw on the pile in case of emergency, and Leona Woods, the only woman on the team, calling out the count from the boron trifluoride neutron detector she had devised in a loud voice, George Weil started withdrawing the rods 6 inches at a time.

As the rods were pulled out, the counts increased as the neutron population provided by the radium source rose through subcritical multiplication. After 25 minutes, with the detector scale having to be adjusted to deal with the rapidly rising count, Fermi announced that the reactor had gone critical. Three minutes later, the exponentially rising count exceeded the safety limit, and Fermi ordered the rods back in, and the experiment shut down nicely.

Compton, who was present representing Roosevelt's science advisory group, picked up the phone and called James Conant in Washington. "The Italian navigator had landed in the New World," he said.

The comparison with Columbus is apt. For just as the world-historic accomplishment of finding America became obvious as soon as Columbus had done it, so Fermi's simple technique for creating a critical nuclear reactor is now clear to all. I strongly recommend it as the model for your first home-built unit. You can try your hand at more complex systems later. (See chapter 7 for instructions.)

Paper cups filled with chianti were passed out for the team to toast their achievement. There was dead silence for 30 seconds. Woods was the first to speak. "I hope we were the first to do it," she said.

Fortunately, they were. The Nazis did indeed have their own atom bomb program, but Werner Heisenberg, who was leading its theoretical division, did the same criticality calculation that

Fermi had done, but got the answer wrong. As far as Heisenberg was concerned, if he was going to build a critical reactor using natural uranium fuel, graphite moderator would not do. He wanted something better, and hydrogen-2, aka heavy hydrogen, $^2H$, or deuterium, with its ultra-low neutron absorption cross section was uniquely suitable.

Light water wouldn't work; heavy water certainly would. But doing the required isotope separation to refine deuterium from natural hydrogen to make heavy water requires special facilities and lots of power, and there was only one place where both of these were available: the Norwegian hydroelectric power plant at Vemork. So the Nazis expropriated it. But the plan went awry when the Norwegian resistance, following Winston's Churchill's polite request to Norway's exiled King Haakon VII, blew the heavy water plant up in 1943. The Germans had, however, by that time already produced enough heavy water to make a bomb factory. In February 1944 they moved to convoy the accumulated supply back to the Reich. The train was far too heavily guarded for the resistance to attack with any hope of success, but they saw that it was headed towards Lake Tinn, where it would need to take a ferry. Examining the train schedule, and that of the lake ferries, the Norwegians determined which ferry it would go on, and at what point the ferry would be traveling over the deepest part of the lake. So, thinking ahead, they planted a time bomb in the ferry bilge, and trusted German punctuality to do the rest. It did. *Und das Projekt war kaputt.*[4]

The Manhattan Project, however, now had not one, but two clear paths forward to success. It took both. In addition to the massive U-235 isotope separation facilities at Oak Ridge, a large graphite-moderated reactor based on the Chicago Pile was set up in Hanford Washington. With a power rating of 100 MWt – a million times that of Fermi's test unit – it would use its massive neutron flux to transmute U-238 into Pu-239, which could then be chemically separated from the uranium to make fuel for fast-fission plutonium bombs. With the help of Woods, who solved the problem of "poisoning" of the chain reaction by high absorption cross section xenon-135 gas fission products, it soon was producing plutonium in quantity.

By summer 1945, enough fissile material had been accumulated to make three bombs. On July 16, in the desert near Alamogordo New Mexico, Allison threw the switch to test the first one. The test, dubbed Trinity, was successful beyond all expectations. A single bomb produced an explosive yield

**Fig. 5.5** On the fourth anniversary of the team's success, December 2, 1946, members of the CP-1 team gathered at the University of Chicago. Back row, from left: Norman Hilberry, Samuel Allison, Thomas Brill, Robert Nobles, Warren Nyer, and Marvin Wilkening. Middle row: Harold Agnew, William Sturm, Harold Lichtenberger, Leona Woods and Leo Szilard. Front row: Enrico Fermi, Walter Zinn, Albert Wattenberg and Herbert L. Anderson. (Credit: Los Alamos National Laboratory)

equal to thousands of tons of dynamite. With the war still raging across Asia and the Pacific costing tens of thousands of American, British, Australian, Kiwi, Filipino, Polynesian, Indonesian, Vietnamese, Indian, Chinese, and Japanese lives every day, and an even more intense bloodbath looming with the impending invasion of Japan, President Harry Truman made the decision to put the new super weapon to use. On August 6 Hiroshima was bombed, then Nagasaki on August 9. Japan's dead-ender militarist leaders had wanted to fight on to national Hari Kari, but in the face of such overwhelming power they had no case. Japan sued for peace immediately, and on August 15, Emperor Hirohito announced the surrender to the Japanese people.

The bombing of Hiroshima and Nagasaki has been controversial ever since. On viewing the Trinity test, J. Robert Oppenheimer, the refined scientific leader of the Manhattan Project, is reported to have morosely quoted the Bhagada Vita, "I am become death, the destroyer of worlds."[5] But for those whose lives were closer to the brutal realities of the war, there was no doubt. As Leona Woods commented years later, "I certainly do recall how I felt when the atomic bombs were used. My brother-in-law was captain of the first minesweeper scheduled into Sasebo Harbor. My brother was a Marine, with a flame thrower on Okinawa. I'm sure these people would not have lasted in an invasion. It was pretty clear the war would continue, with half a million of our fighting men dead not to say how many Japanese....I have no regrets. I think we did right, and we couldn't have done it differently."[6]

As for me, I wasn't around yet, but at the time of the bombing, my dad was a US Army sergeant in a unit scheduled to take part in the invasion of Japan. So, thank you Harry.

Americans rejoiced. The bloodiest war in human history was over.

**Fig. 5.6** Oak Ridge, Tennessee, August 14, 1945. Americans everywhere celebrated the end of the war. But the workers at the Manhattan Project's Oak Ridge facility had extra reason to cheer. They had helped strike the knockout blow. (Credits: Wikipedia Commons)

# CHAPTER 6
# ATOMS FOR SUBS, ATOMS FOR PEACE

**A NEW POWER** of unimaginable dimensions was loose in the world, but was there anything that could be done with it beyond making superbombs? There were innumerable possibilities that could be thought of offhand, but if nuclear researchers were to be diverted from bomb making, any additional applications needed to be of serious military interest. Fortunately, there was one.

At the end of World War II, the United States found itself with global military and economic commitments which could only be supported by maintaining control over the seas. Yet the advent of the atomic bomb had made large fleets of aircraft carriers, battleships, cruisers, and other surface combatants vulnerable and obsolete. Submarines were a vital alternative, but the submarines of World War II were not true submarines, because they had to spend most of their time running on the surface powered by air-breathing diesel engines. Powered underwater at low speed by their electric storage batteries, they had a maximum submerged endurance of 24 to 48 hours. The Navy needed a power source that could operate indefinitely below the water, with an effectiveness and reliability comparable to that which diesel engines provided surface vessels.

Nuclear power seemed like a hopeful solution, and so in 1946, the U.S. Navy's Bureau of Ships sent a team of engineers to the Atomic Energy Commission's laboratories at Oak Ridge, Tennessee to study nuclear technology and its possible naval applications. Because of the uncooperative attitude of the AEC bureaucracy, Bureau of Ships Chief Earle Mills chose his most abrasive officer to head the team, Captain H. G. Rickover. A hard-driving Polish-Jewish immigrant who had graduated from the U.S. Naval Academy at Annapolis, Rickover had served as head of the Electrical section of the Bureau of Ships during World War II. There he had earned a reputation not only as a top engineer, but also as a man who would crush any bureaucratic or procedural obstacles that stood in the way of getting a vital task done.[1]

Deploying his Navy team at Oak Ridge like a search-and seize task force to ferret out information, Rickover came to the conclusion that developing a nuclear power reactor was no longer a theoretical question but simply an engineering problem. He reported back to the Navy that the development and construction of a nuclear reactor for submarine propulsion should be made a number-one priority, and was quickly able to win over Bureau of Ships Chief Mills and Chief of Naval Operations Admiral Chester Nimitz to his viewpoint.

As expected, the AEC bureaucracy resisted any diversion of efforts from bomb making, or any dilution of its authority over all things nuclear. But the buildup of a large Soviet submarine force by 1948, followed by the detonation by the Soviets of an atomic bomb in 1949, greatly strengthened the urgency of Rickover's case. With that wind at his back, Rickover was able to engineer a deal, setting up a naval reactor program under joint Navy-AEC auspices, with himself as project manager for both agencies.

Rickover asked industry for ideas, and Westinghouse came up with a concept that he liked a lot. That concept generated power by making steam, using a loop of pressurized water cycling through the reactor to both take out heat from the reactor to the steam generator, and as the moderator for the reactor itself. By this time, plenty of highly enriched U-235 was available, and using enriched fuel, ordinary "light" water would work fine as a moderator.

But here's the thing. While the water moderator/reactor coolant was pressurized, allowing it to get to temperatures over 350 C before it would boil, still if it got *too* hot, it would boil regardless. If the reactor power level started to rise beyond the capacity of the cooling water to keep its temperature below the high-pressure boiling point, that would happen. In fact, there would be no way to stop it from happening. But if the moderator water boils, it will, from the neutron's point of view, start to have holes in it and *it will fail to do its job as a moderator*. Without adequate moderation, the neutrons start missing, and the growth of the reaction stops. It's negative feedback – like having a stove which automatically turns itself down if the water starts boiling too hard, except in a reactor built in this way, the negative feedback doesn't come from an artificial control system that can fail, but from laws of physics that can never fail. This is why a water-moderated reactor can never explode like a nuclear bomb, or even have a runaway power excursion. If it gets too hot, the water boils more than it should, and the reaction stops.

This is a central example of a broader principle. The exponential nature of neutron multiplication might appear to make a nuclear reactor very hard to control, except that when neutrons multiply the fission energy they release instantly heats everything up around them. This very action of heating up the surrounding material can cause the neutrons' batting average to decline. For example, as materials heat up they expand, which causes their nuclei to become more loosely packed, increasing the probability that fast-moving newborn neutrons will miss their targets before flying out of the fissile region. But when water boils, its density is reduced not by a few percent, as occurs during solid material temperature expansion, but by orders of magnitude. If water is your moderator, the negative feedback from overheating is so powerful it can stop the growth of a neutron population dead in its tracks.

Of course, if the reaction stops, temperatures will drop, boiling will decrease, and the reaction will grow back, and this will all happen very fast. So, what really happens is that once boiling reaches the maximum level at which a critical reaction can be sustained, the power will level out. At that point, if you want to increase the power, you have to pump more water. If you want to drop the power level, just pump less. Since you control the pumps, you control the power. It's that simple. Rickover loved it.

On July 15, 1949, the contract was signed, and the project that was to lead to the world's first nuclear submarine, was underway. She would be named the *Nautilus*, after fabulous submarine in Jules Verne's novel *Twenty Thousand Leagues Under the Sea*, and the US Navy's own USS *Nautilus* SF-9/ SS-168, which served with great distinction in World War II.

Rickover, however, was skating on thin ice. His ethnic background made him an outsider to the Navy brass WASP fraternity, and his personality was anything but smooth. As a result, he was repeatedly denied promotion from Captain to Rear Admiral. Furthermore, just as the prewar battleship admirals had dismissed aircraft carriers, the carrier admirals who dominated the highest echelons of the postwar Navy were less than enthusiastic about a technical development that threatened to make submarines the true capital ships of future naval warfare.

So Rickover needed to move fast. To maximize the rate of development of the project, he decided to avoid building many scaled-down prototype reactors. Instead, only one test reactor would be built, the Mark I, which would be identical to the

Mark II reactor that would eventually be installed in the *Nautilus*, whose hull was already under construction. The path of building the Mark I spread out over a large floor for easy access was rejected; instead, it was installed in a submarine hull built into the Mark I test site in Idaho, surrounded by a huge tank of water so that all the radiation reflection experienced by a submerged submarine could be simulated. And rather than cool the reactor room with open air, air conditioning was built into the Mark I, since that was the way the *Nautilus* reactor room would have to be cooled. The Mark I components were placed in an old submarine and depth charged in Chesapeake Bay; those that could not take the shock were redesigned.[2]

In all respects, the operative design slogan was "Mark I equals Mark II." If Mark I functioned adequately, so would the *Nautilus*. By the end of May 1953, the Mark I reactor was completed.

Yet, with the project on the brink of success, the change of presidential administrations following the 1952 election put it in extreme danger. In February 1953, Robert B. Anderson, a Texan with close political ties to Texas oil barons Sid Richardson and Clint Murchison, FBI Director J. Edgar Hoover, and anticommunist witch hunter Senator Joe McCarthy, was appointed Secretary of the Navy. Rickover was certainly no red, but he was an FDR-Truman Democrat, precisely the faction McCarthy had targeted for destruction. Furthermore, he was championing a technology that directly threatened the oil industry. His enemies demanded his retirement by no later than June 30, 1953. If the project did not achieve success immediately, it would be shut down.

There was no time to be lost. Following a series of preliminary tests, Rickover brought the Mark I to full power on June 25. It ran smoothly for 24 hours, so the officers on the site decided to end the test. Rickover overruled them. He ordered that charts be brought into the control room and a simulated great circle course to Ireland be plotted. No submarine had ever traveled more than 20 miles submerged at full speed before. At the 60th hour, the nuclear instrumentation became erratic; then problems developed with the reactor cooling pumps. At the 65th hour, a condenser tube failed, and steam pressure fell off rapidly. But Rickover refused all requests by Navy and Westinghouse officials to terminate the test. "If the plant has a limitation so serious," he said, "now is the time to find out. I accept full responsibility for any casualty". Repairs on the faulty equipment were undertaken with the reactor running at full power. At the end of 100 tense hours, the position marker on the chart reached Fastnet. A nuclear-powered submarine had, in effect, steamed non-stop across the Atlantic without surfacing.[3]

When the test ended, Rickover turned to his right-hand man, Ed Kintner. "We had to do it today," Rickover said. "They never would have given us another chance."[4]

Six months later the *Nautilus* (SSN-571) was launched, and within a year it was breaking all records. In April 1955, the *Nautilus* traveled submerged from New York to Puerto Rico, 10 times the distance any submarine had ever traveled under water. In war games held in August of that year, the *Nautilus* demolished (in simulation) an anti-submarine task force consisting of an aircraft carrier and several destroyers; its high speed and unlimited submerged endurance made it almost invulnerable. Congress immediately decided to order six more nuclear submarines.

**Fig. 6.1** Rickover on the Nautilus
(Credit: US Navy)

**Fig. 6.2** The Nautilus demonstrated the astounding new capabilities offered by nuclear power. In 1958 she sailed under the North Pole. In 1959, the Skate (SSN-578) surfaced there. (Credit: US Navy)

The nuclear navy was launched, and with it, the nuclear industry. But Rickover quickly realized that a nuclear development program of the dimensions he envisioned could not succeed by raiding manpower from the precious few nuclear engineers and scientists available. In 1949 he sent aides to MIT and Oak Ridge National Laboratories to persuade those institutions to set up schools of nuclear engineering, and simultaneously initiated a series of courses for his Washington staff and submarine crews in reactor theory, physics, mathematics, nuclear engineering, and naval architecture. The net result was creation not of trained personnel in the ordinary sense of the term but of topnotch engineering cadre, who could not only operate a nuclear reactor, but design and build one. By the time Rickover retired in 1979, 7,000 officers and 40,000 enlisted men had graduated from this curriculum. Today these men (and now women, as well) represent the core of the engineering and technical cadre of the American nuclear industry. Indeed, the majority of all U.S. nuclear plant operators are nuclear navy graduates, a testimony to his program.

## ATOMS FOR PEACE

In fact, launching a commercial nuclear industry was exactly what Rickover had in mind. When budget cutters in the Eisenhower administration, working with traditionalist elements in the Navy, managed to kill his program for a nuclear-powered aircraft carrier, Rickover fought back by proposing that the carrier reactor program, already under preliminary development by Westinghouse, be continued under AEC sponsorship as a program to develop a civilian atomic energy plant.

This idea found support within the industry and the AEC, but was adamantly opposed by Navy Secretary Robert B. Anderson. Anderson, the oil industry ally who had just rejected Rickover's plan for a naval nuclear carrier, now said that the Navy could have nothing to do with commercial nuclear power since it was strictly a civilian enterprise. But news from the Soviet Union once again strengthened Rickover's hand. In August 1953, the Soviets exploded the world's first practical hydrogen bomb. Rickover's ally on the AEC, Thomas Murray, took advantage of the occasion to write President Eisenhower, urging that the United States could carry out a major coup by answering the Soviet development with an announcement of a full-scale U.S. civilian nuclear energy program. Atoms for peace would be the American answer to Soviet atoms for war.

While the administration was mulling over this proposal, Murray acted, delivering a historic speech in Chicago on Oct. 22, 1953. The United States must take steps to· develop nuclear energy for the electric-power-hungry countries of the world,

Murray said, or else the nation would not be able to count on them for the uranium ore upon which U.S. nuclear weapons and national security depended. Finally, on December 8, President Eisenhower delivered his famous "Atoms for Peace" speech to the United Nations, committing the United States to lead the way in the peaceful exploitation of nuclear power for all mankind.

"It is not enough to take this weapon out of the hands of the soldiers, Eisenhower said. "It must be put into the hands of those who will know how to strip its military casing and adapt it to the arts of peace.

"The United States knows that if the fearful trend of atomic military build-up can be reversed, this greatest of destructive forces can be developed into a great boon, for the benefit of all mankind. The United States knows that peaceful power from atomic energy is no dream of the future. The capability, already proved, is here today. Who can doubt that, if the entire body of the world's scientists and engineers had adequate amounts of fissionable material with which to test and develop their ideas, this capability would rapidly be transformed into universal, efficient and economic usage?"

"To the making of these fateful decisions," the president concluded, "the United States pledges before you, and therefore before the world, its determination to help solve the fearful atomic dilemma – to devote its entire heart and mind to finding the way by which the miraculous inventiveness of man shall not be dedicated to his death, but consecrated to his life."[5]

**Fig. 6.3** US 3 cent postage stamp issued in 1955 celebrates Atoms for Peace (Credit: US Atomic Energy Commission)

With the launching of the Atoms for Peace program, the development of a civilian nuclear reactor became a national priority, and the responsibility for getting the job done could only be given to Rickover and his team at the Naval Reactors Branch. A group was soon assembled that consisted of Rickover's Navy team, Westinghouse, Stone and Webster, Burns and Roe, and the Duquesne Power and Light Company of Pittsburgh. Contracts were signed; and on Sept. 6, 1954, President Eisenhower used a radioactive wand to activate the bulldozer that broke ground for the construction of the nation 's first nuclear power plant at Shippingport, Pennsylvania.

Rickover's team worked closely with Westinghouse R&D people, and his staff drove the pace of construction furiously. Despite strikes and steel shortages, the plant was completed in just three years (today it can take sixteen), and by Dec. 23, 1957 it was generating power at full capacity.

Although small by current standards (60 megawatts), the Shippingport plant had an enormous impact on the development of civilian nuclear technology. It took its design point of departure from Rickover's light water moderated submarine reactors. These used 93 percent enriched uranium-235 for fuel. However you can make a light water moderated reactor critical using even 3 percent enriched U-235, which is much cheaper. Furthermore, unlike a naval reactor, a land-based civilian reactor can be refueled, whenever convenient, and is not size-constrained, so it has no need for either the indefinite durations operation or compact size required by naval reactors. Finally, and most importantly, while 3 percent enriched U-235 can be made critical using slow neutrons, it cannot be made critical in an unmoderated system. It is therefore useless as bomb fuel, and immune to diversion for such purposes — a criterion the Rickover insisted on as a requirement for a civilian nuclear reactor. Designed accordingly, the Shippingport system had no direct

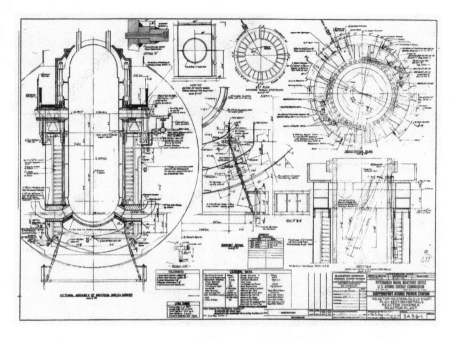

**Fig. 6.4** To support the Atoms for Peace program, the AEC broke with Cold War secrecy and published the design of the Shippingport nuclear power plant. (Credit: US Atomic Energy Commission)

military applications. So, consistent with the intent behind the Atoms for Peace program, its design was unclassified.

Hundreds of engineers from around the world attended seminars on the Shippingport power plant given by the Naval Reactors Branch, Westinghouse, and Duquesne during 1954-55, and Westinghouse made available thousands of technical reports on every aspect of the project. Following its opening, it continued to serve as a school in reactor technology for hundreds of engineers.

As a result, the pressurized water reactor design became the model for more than three fourths of all civilian nuclear reactors produced worldwide ever since.

For many years two framed quotes hung on the walls of Rickover's Washington office. One was from Shakespeare's *Measure for Measure*:

> "Our doubts are traitors
> And make us lose the good we oft might win,
> By fearing to attempt."

The other was from the Bible:

> "Where there is no vision, the people perish."

# CHAPTER 7

# HOW TO BUILD A NUCLEAR REACTOR

**NUCLEAR POWER CAN** provide infinite energy for an unlimited human future, while avoiding the droughts, floods, ice ages, heat waves, ocean acidification or other Great Filter events said to be responsible for the extinction of all those extraterrestrial civilizations throughout the galaxy who would otherwise be communicating glibly with us today.

But it is something else too. It's a great way to make money. Now, if you've built your nuclear pile as I instructed you to do in Chapter 5, you should have a good hands-on knowledge of the basic nuclear physics involved. But you'll need to know more nuclear engineering to build a practical reactor that can make you some serious moolah. You'll also need more cash, because it will cost you a lot more to build a gigawatt scale reactor than a simple atomic pile. But don't worry. You can get all the cash you need by starting a dot com or crypto currency company.

Of course, if you did that you could get rich without building any reactors, but what would be the fun of that? The key to a good life is not just making money, but doing it while accomplishing something of world historical benefit to humankind. That way you can feel happy about your riches, and will have a ready answer when your grandchildren ask you "What did you do to save humanity from the Great Filter?" So that should be your plan.

But even if you are one of those greedy people who just wants to get rich without building any nuclear reactors, it's important to know how they are built so you can understand the issues involved and play a knowledgeable and constructive role in the important public policy debates concerning their use.

So, let's have a look at how you might build a Pressurized Water Reactor (PWR.) If you want to get into selling nuclear power stations – or just understand the nuclear business – these are a smart place to start, as they are the most popular type by far. Something like seventy percent of all reactors worldwide are PWRs, with another twenty percent of so being related types, such as the Canadian CANDU and the Boiling Water Reactor (BWR).

The major components of a PWR power station are shown in Fig. 7.1

If you look at Fig. 7.1 you will see the main pieces you will need to build your PWR. The largest of these is the containment

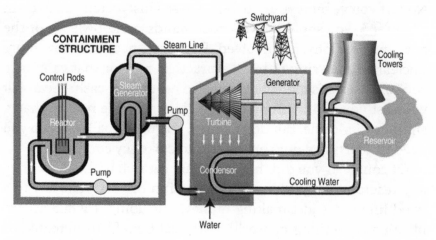

**Fig. 7.1** A Pressurized Water Reactor power station. (Credit: US Department of Energy)

building, into which you will put the steam generator, and the reactor pressure vessel, which contains the reactor itself. A line from the steam generator takes the steam out of the containment building to a drive a steam turbine, which generates the electric power with the help of a cooling system, just as it would if it were being fed steam made by any other heat source. The turbine, its cooling system, the electrical transformers and all the rest of the junk outside the containment building are all garden variety off the shelf power company hardware. You can pick it all up on eBay. Even better, just tell the utility that you expect them to have it in place already, so you can plug and play. Why should you need to trouble yourself with some standard 19th century technology? It's the containment building and everything inside of it that will require your special expertise. So let's focus on that.

All American civilian power reactors, by law, require a containment building. Military reactors, including those on submarines and ships that dock regularly in port cities do not. That is because laws are for the little people. (It may also be true that putting a nuclear aircraft carrier inside a containment building could hamper flight operations, but such considerations carry no weight with the Nuclear Regulatory Commission.) It's completely unfair, I know, but if you want to get into the nuclear power business in any western country, you'll just have to humor the authorities on this point and go along. The specifications for the containment building are very tough. They need to be able to withstand direct impact by airliners crashing into them at full speed. The Atomic Energy Commission thought this requirement up decades before the 9-11 hijackers ever rolled off of their fruit truck. It may seem to be very difficult to engineer buildings this strong, but in fact the necessary technology of steel reinforced concrete was mastered by the Germans in World War II, who built submarine pens on the coast of France that resisted damage from repeated ferocious allied air attacks using blockbuster bombs.[1] (They might not have withstood a nuclear attack, but if anyone drops a nuke on your plant the people in the neighborhood are going to have a lot more to worry about than you.) Containment buildings are based on this technology, greatly improved, employing steel reinforced concrete walls 1.3 meters (4 ft) thick, with a base 2.6 meters (8 ft) thick.

So, while having a containment building may be the law, it's really a good idea too. Because let's say there is a terrorist, or a competitor, or an unsatisfied customer or ex-spouse who wants to get you by crashing an airliner into your nuclear power plant. If you have a containment building, you can just smile and tell them to bring it on. Of course, since you are a reader of this book, you may well be so lovable that there are no such people in your life. Even so, you should not try to save money by omitting a containment building from your plant's design. The Nuclear Regulatory Commission sends out inspectors every now and then to determine if you are following regulations. They are not the sharpest people you will ever meet, but having a containment building is on their checklist. If it's not there they might notice.

Once you have a containment building, you will need to get a steam generator and a pressure vessel to put inside of it. The steam generator uses very hot (345 C) pressurized water from the reactor to make steam to feed to the turbine. This is not rocket science. It's the sort of thing that has been around ever since they started making steam turbines to drive Dreadnoughts. You can probably pick one up cheap in a Royal Navy scrapyard or similar high-quality used parts supplier. The pressure vessel, however, is a specialty hardware item, because it needs to be able to take pressures of 2280 psi (155 atmospheres), with a good safety margin to boot. It needs to take 2280 psi because that's the pressure you need to keep water from boiling at 345 C, which is right where you want it. The pressure level determines the maximum temperature of a water-moderated reactor, because the boiling point of water is determined by its pressure, and if the water starts boiling too vigorously, voids will appear in the moderator, automatically curtailing the chain reaction. This is a key safety feature of all water moderated systems. You could operate at a lower pressure, and also make more allowance for some boiling, as the boiling water reactors (BWRs, 300 C, 1000 psi) do. This would allow you to use a thinner and cheaper pressure vessel. But, unless you were willing to reduce your power generation efficiency by using low-temperature steam, you would have to drop the secondary loop steam generator and take water that has been in the reactor all the way through the steam turbines. That's exactly what BWRs do, which makes them simpler than PWRS. But small leaks of radioactive products from the fuel into this water can make it slightly radioactive, which means that it can make the steam turbine radioactive too, unless you have a really good filter system to prevent it. This, however, is not possible in a PWR, because the steam that goes to the turbine system comes from a secondary loop, which has never been in the reactor, so it is guaranteed to be completely clean. That's why PWRS are so much more popular than BWRs — with utilities favoring them four to one in the most recent poll — and why I recommend PWRs too. Three hundred happy PWR owners can't all be wrong.

So going with a good strong PWR pressure vessel is the right way to proceed. If you choose a 4-meter diameter vessel, you'll need to make the walls 20 cm (about 8") thick. That's a lot of metal, but ordinary cheap carbon steel will do the job, so don't let any specialty metal salesmen con you into buying zirconium or titanium or other fancy stuff. You will need to clad the carbon steel with stainless to prevent corrosion, but that needn't cost too much.

Excellent pressure vessels can be purchased from the Westinghouse company. They are expensive, however, so ask about

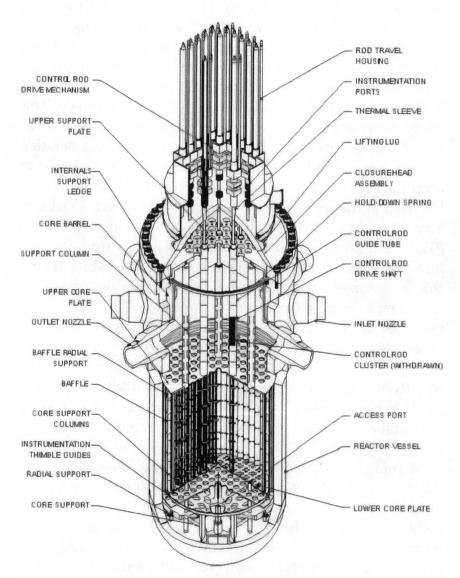

CONTROL ROD
DRIVE MECHANISM

UPPER SUPPORT
PLATE

INTERNALS
SUPPORT
LEDGE

CORE BARREL

SUPPORT COLUMN

UPPER CORE
PLATE

OUTLET NOZZLE

BAFFLE RADIAL
SUPPORT

BAFFLE

CORE SUPPORT
COLUMNS

INSTRUMENTATION
THIMBLE GUIDES

RADIAL SUPPORT

CORE SUPPORT

ROD TRAVEL
HOUSING

INSTRUMENTATION
PORTS

THERMAL SLEEVE

LIFTING LUG

CLOSUREHEAD
ASSEMBLY

HOLD-DOWN SPRING

CONTROLROD
GUIDE TUBE

CONTROLROD
DRIVE SHAFT

INLET NOZZLE

CONTROLROD
CLUSTER (WITHDRAWN)

ACCESS PORT

REACTOR VESSEL

LOWER CORE PLATE

**Fig 7.2** PWR pressure vessel with bolt on top and control rods. The fuel assemblies go inside. (Credit: US Department of Energy)

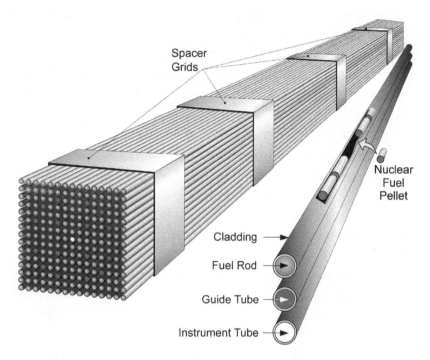

**Fig. 7.3.** A PWR fuel assembly consists of a bundle of fuel rods, which are tubes containing a line or uranium oxide fuel pellets. Water can flow through the channels between the rods to provide cooling. (Credit: US Department of Energy)

their layaway plan. Or, if you are good at welding, you can just make it yourself. You'll need to make it in two pieces, with the top, containing tubes for the control rods, bolting on using a set of flanges, as shown in Fig. 7.2.

Now that you have a containment building, steam cycle equipment, and a pressure vessel, it's time to put together your fuel assemblies.

A typical fuel assembly consists of a bundle of 200 to 300 fuel rods, with each rod containing a line of little uranium oxide ($UO_2$) fuel pellets, as shown in Fig. 7.3. As you can see, there are open channels between the rods. These provide a path for the cooling water.

The tubes which form the rods are made of metal. Sorry, but you'll need to pay the piper and use zirconium for these. Steel just won't do. This is because, as you can see from the table in Chapter 5, the absorption cross section of iron and its usual alloying materials like nickel and chromium are on the high side, while that of zirconium is much less. It's okay to have them in the pressure vessel, because any neutrons that head out that way are deserting the reaction in any case, so if they get absorbed that's just their tough luck. But the fuel rod tubes are in the heart of the action, any neutron they absorb is a loss to the cause.

You'll need to get *a lot* of zirconium alloy tubes, as you'll require several hundred for each bundle, and at least a hundred bundles in all to make your reactor. But fortunately, quantity discounts are available.

Once your fuel assemblies are all built, you can insert the fuel pellets in them. This is the fun part, because it's the fuel pellets that make all the rest of it go.

Fuel pellets are made of uranium oxide. Quality is essential, and you need to make sure you are getting the right product at time of sale, because most vendors won't take them back after they are used. Using what you learned in high school chemistry, you can easily test your pellets to make sure they are made of pure uranium oxide, but the real issue is to know that the uranium portion contains the right amount of U-235, because that's the stuff that gives your fuel its kick.

If you can get your hands on some heavy water, which is the best moderator there is, you could use natural uranium for fuel. That is how the Canadian CANDU reactors are designed, taking advantage of that country's large hydroelectric driven heavy water production capacity. But unless you can get your hands on the trainload of Nazi heavy water that is still sitting on the sunken ferry on the bottom of Lake Tinn in Norway, you will probably find heavy water moderator too expensive. That means you will need to use light water (i.e. ordinary water) to moderate your reactor. Such a system won't work using natural uranium, which is only 0.7% U-235. You need uranium whose U-235 content is enriched to 3%, or about four times that found in nature. If some dishonest vendor sticks you with pellets made from cheap natural uranium, your reactor will be a dud, and you'll be out a bundle with no return. How can you know what you are getting?

There are several ways to inspect your fuel. If you happen to own a good mass spectrometer, you can just take a sample and compare the number of nuclei with mass 235 with those of mass 238. But if you don't want to bother with that, a simpler way is just to see how radioactive the fuel is. U-235 has a half-life of 703 million years, while U-238 has a half-life of 4.5 billion years. As a result, U-235 is 6.4 times as radioactive as U-238 (The shorter an isotope's half-life, the more radioactive it is, because if an isotope has a shorter half-life, it is doing its thing faster.) So all you really need to do the job is a $100 Geiger counter. If you test a given amount of natural uranium and find it showing 104 counts per minute (99 from the 99.3% U-238 and 5 from the 0.7% U-235), then the same amount of 3% enriched uranium should give 116 counts (97 from the 97% U-238 and 19 from the 3% U-235.) If it does, buy it. If it gives 600 counts, then it's 93% enriched, and you should buy all you can and cut it thirty to one with natural uranium to get excellent fuel at a great discount. Or you can just report the seller to the CIA and ask for a reward. Your choice.

Once you have your fuel pellets, you can load them into the fuel rod assemblies and put the assemblies in the pressure vessel. The next step is to insert some control rods. These can be made of any high neutron absorbing material. Fermi used cadmium metal nailed to wooden boards. That worked fine for him because his Chicago Pile reactor never got hot. Your reactor will heat up a lot, so cadmium on wood would not be your best choice. If you like spending money, you can make excellent control rods from a cadmium-silver-indium alloy. But high boron steel, or boron carbide ($B_4C$) encased in steel will also do the job, and is much cheaper, so that is what I recommend.

After you have inserted the control rods into your reactor, you can fill it with water. IMPORTANT NOTE: DO *NOT* FILL YOUR REACTOR WITH WATER UNTIL **AFTER** THE CONTROL RODS ARE IN. Now start the primary and secondary loop water pumps to make sure they work. These pumps should be operated using electricity from the grid, not from your reactor, because you want them to be able to work even if the reactor is shut down. You should also have at least two standby diesel generators on site, with each by itself able to generate enough power to operate the reactor pumps should the grid go down, and plenty of fuel stored nearby for each.

With these preparations in place, you can now start pulling out the control rods. Just as it did with your atomic pile, the neutron population will start to increase. When the rods have been withdrawn far enough to allow $k$, the per capita birthrate of your neutrons to reach 0.9, the neutron population of your reactor will grow to ten times the rate provided by your neutron source. Pull the rods out a bit further, and $k$ will grow to 0.99, raising your neutron population to 100 times the source. That will still generate negligible power. But pull the rods out just a bit more, to raise k to 1.0001. Now the neutrons will start multiplying on their own, increasing by 0.01% every millisecond-long slow neutron generation. In about 20 seconds their number will increase tenfold, and they will keep increasing tenfold every 20 seconds, multiplying a millionfold in two minutes. Soon they will start to generate serious power, heating the water up. But don't worry; as soon as the water reaches 345 C, the multiplication will stop as the smallest amount of boiling takes away the moderation the reaction needs to grow, and cuts $k$ down to 1.0000000. So, with your chain reaction running steady at its design level, start making steam and spin your turbine.

Your nuke is in business.

**Fig. 7.4.** When you are done, your nuclear power station should look like this. The Sequoya Nuclear Plant, Soddy-Daisy, Tennessee uses two Westinghouse PWRs to generate 2440 MW of electricity. (Credit: Wikipedia Commons)

# CHAPTER 8

# IS NUCLEAR POWER SAFE?

**A LOT OF** fuss has been raised about nuclear power plants. Some say they emit cancer-causing radiation, that there is no way to dispose of the waste they produce, that they are prone to catastrophic accidents, and could even be made to explode like bombs!

These are serious charges, and now that you have started building nuclear reactors, or at least learned enough about them to be able to build a few when you have the time, you are well-qualified to look into them.

So, let's investigate the alleged dangers, starting with routine emissions, then waste disposal, then accidents. We'll save bomb-making for last, since it's the most exciting part of the discussion, and I want to hold everybody's interest till the end.

## ROUTINE NUCLEAR POWER PLANT RADIOLOGICAL EMISSIONS

Americans measure radiation doses in units called rems, or, more often, millirems (abbreviated mrem), which are thousandths of a rem. Europeans use units called Sieverts. That is because they are snoots and want to show off by using bigger units. (One Sievert equals 100 rems.) While high doses of radiation delivered over short periods of time can cause radiation poisoning or cancer, there is, according to the U.S. Nuclear Regulatory Commission, "no data to establish unequivocally the occurrence of cancer following exposure to low doses and dose rates—below 10,000 mrem."[1] Despite this scientific fact, the NRC and other international regulatory authorities insist on using what is known as the "Linear No Threshold" (LNT) method for assessing risk.

The LNT methodology was developed by Hermann J. Muller. While he was a real scientist who won the Nobel Prize for his work on fruit fly genetics in 1946, Muller was also a fanatical eugenicist and a communist with connections at the highest level of the Communist International. Indeed, he was personally acquainted with Soviet leader Josef Stalin. This got him into some trouble when he vigorously urged Stalin to make use of the unique advantages offered by the dictatorship of the proletariat to control the reproduction of the Soviet population. The right to reproduce, advised Muller, should be limited to the eugenically endowed top one percent of Soviet men, thereby accelerating the improvement of humanity under scientific socialism. The potential political downside of such a program did not escape Stalin's observation, and concluding that Muller was either nuts or a capitalist provocateur, ordered him executed. However, warned by friends in the NKVD secret police, Muller escaped the country in advance of the arrest squad and joined the International Brigades in Spain before ultimately coming to the USA. There he helped found the Population Council (a Malthusian organization dedicated to reducing world population) and the Pugwash movement attempting to arrange for collaboration between Anglo-American and Soviet elites.[2]

According to Muller's LNT methodology, a low dose of radiation carries a proportional fraction of the risk of a larger dose. So according to LNT theory, since a 1000 rem dose represents a 100 percent risk of a death, then a 100 mrem does should carry a 0.01% risk. If this were true, then one person would die for every 10,000 people exposed to 100 mrem. Since there are 330 million Americans and they already receive an average of 270 mrem per year, this would work out to 90,000 Americans dying every year from background radiation, a result with no relationship to reality. Fundamentally, the fallacy of the LNT theory is the same as concluding that since drinking 100 glasses of wine in an hour would kill you, drinking one glass represents a one percent risk of death. It's quite absurd, and the regulators know it. But we are talking government regulators here, so naturally they use it anyway. That said, let's look at the data.

The annual radiation doses that each American can expect to receive from both natural and artificial radiation sources are given in Table 8.1.

## TABLE 8.1 RADIATION DOSES FROM NATURAL AND ARTIFICIAL SOURCES[3]

| | |
|---|---|
| Blood | 20 mrem/year |
| Building Materials | 35 mrem/year |
| Food | 25 mrem/year |
| Soil | 11 mrem/year |
| Cosmic Rays (sea level) | 35 mrem/year |
| Cosmic Rays (Denver altitude) | 70 mrem/year |
| Dental X-Ray | 10 mrem |
| CT Scan (head & body) | 1100 mrem |
| Air travel (New York to LA round trip) | 5 mrem |
| Nuclear power plant (limit, at property line) | 5 mrem/year |
| **Nuclear power plants (dose to general public)** | **0.01 mrem/year** |
| Average annual dose (general public) | 270 mrem/year |

Examining Table 8.1, we see that the amount of radiation dose that the public receives from nuclear power plants is insignificant compared to what they receive from their own blood (which contains radioactive potassium-40), from the homes they live in, from the food they eat (watch out for bananas), from the medical care and air travel they enjoy, from the planet on which they reside, and from the universe in which their planet resides.

In fact, far from increasing the radiological exposure of the public, nuclear power plants reduce it. Coal contains radioactive constituents. Worldwide, coal-fired electricity stations release some 30,000 tons of radioactive radon, uranium and thorium into the atmosphere every year. They also emit millions of tons of highly toxic chemical ash containing mercury, arsenic, selenium, not to mention over ten billion tons of $CO_2$ per year. In fact, along with 10 million tons of $CO_2$, a *single* 1000 MWe coal-fired power plant annually produces 200,000 tons of ash, and in addition to several hundreds of tons of mercury and other chemical poisons, sends some 27 tons of radiative material – half radon, the other half uranium and thorium – right up the stack. In fact, the amount of uranium and thorium emitted to the environment as pollution by coal-fired power plants would be more than enough to fuel every nuclear power plant in the country, which could produce equivalent power without any of the $CO_2$, toxic gas, or radiological emissions.

Natural gas is much cleaner than coal, but it contains radioactive radon. Not much, to be sure, typically about 0.03 microcuries per cubic meter. But that adds up. A 1000 MWe natural gas power plant sends about 8 Curies of radon into the environment *every month*. That's just about the same as what the Three Mile Island nuclear power plant let loose just once – during its world-famous meltdown in March 1979![4]

Now, I am not saying that coal and natural gas power plants should be shut down because of their radioactive releases. But since they routinely emit more radiation than not only well-functioning nuclear power plants, but a nuke during the worst reactor accident in US history, the fervor of the authorities and the activists in this area would appear to be wildly misplaced.

So if your virtue-signaling old college friends are giving you a hard time about the radiation from your nuclear power plant, tell them that you are not the problem, they are the problem. Are they trying to irradiate the world with fossil fuels? And what are they doing about bananas?

## NUCLEAR WASTE DISPOSAL

One of the strangest arguments against nuclear power is the claim that there is nothing that can be done with the waste. In fact, it is the compact nature of the limited waste produced by nuclear energy that makes it uniquely attractive. A single 1000 MWe coal-fired power plant produces about 600 tons of highly toxic waste daily, which is more than the entire American nuclear industry does in a year. The portion of this that is not simply sent up the stack piles up near the power plant, or is dumped somewhere else, and eventually finds its way into the biosphere. Despite the clear, non-hypothetical consequences of this large-scale toxic pollution, no one is even talking about establishing a waste isolation facility for this material, because it is not remotely possible. In contrast to such an intractable problem, the disposal of nuclear waste is trivial.

It must be said: The hazards of nuclear waste disposal have been exaggerated by environmentalists, with the *openly stated purpose* of seeking to create a showstopper for the nuclear industry. They claim to be interested in public safety and ecological preservation. Neither claim is supportable. By protecting fossil power from nuclear power, the anti-nukes are perpetuating environmental devastation and harming public health. Regarding public safety they are even worse. Indeed, it must perplex the rational mind that anyone can agitate, litigate, and argue with a straight face that it is better that nuclear waste be stored in hundreds of cooling ponds adjacent to reactors located near metropolitan areas all across the country than that they be gathered up and laid to rest in a government-supervised depository in a location far removed from civilization. Yet that is what they do.

There are two excellent places to store nuclear waste: under the ocean bottom or under the desert. The US Department of Energy has opted for the desert, but the ocean solution is much simpler and cheaper. Let's talk about that first.

The way to dispose of nuclear waste at sea works as follows: First, you glassify the waste into a water-insoluble form. Then you put it in stainless steel cans, take it out in a ship, and drop it into the mid-ocean, directly above sub-seabed sediments that have been, and will be, geologically stable for tens of millions of years. Falling down through several thousand meters of water, your canisters will reach velocities that will allow them to bury themselves deep under the mud. After that, your waste is not going anywhere, and no one will ever be able to get their hands on it. Furthermore, no nomads roving the Earth after the next ice age eliminates all records of our civilization will ever be harmed as a result of accidentally stumbling upon it. (I mention this latter point because protection of the public for the next 10,000 years, under all contingencies, has been made a Department of Energy nuclear waste repository requirement.)

This solution has been well-known for years.[5] Unfortunately, it has been shunned by Energy Department bureaucrats who – despite their concern for post Ice Age wandering cannibals – seemingly prefer a large land-based facility because it involves a much bigger budget, as well as by environmentalists who wish to prevent the problem of nuclear waste disposal from being solved. Thus, in the 1980s, the DOE looked the other way and allowed Greenpeace to pressure the London

Dumping Convention into banning sub seabed disposal of nuclear waste. That ban expires in 2025. If world leaders are in any way serious about finding an alternative to fossil fuels to meet the energy needs of modern society, they will see that the ban is not renewed.

If, however, the ban is renewed, the Department of Energy's plan to put the waste under Yucca Mountain in the Nevada desert remains an alternative. While wildly over-priced, the plan has been exhaustively and thoroughly vetted, and it meets even the most stringent standards of public safety. (To save the post ice age savages, the public dosage would be required to stay below 15 millirems of radiation per year for at least 10,000 years.[6] The best estimates, though, show that the average public dosage would be far, far less: under 0.0001 mrem/year for 10,000 years.[7])

However, regardless of the fact that the project has been thoroughly analyzed – the site has been called "the most studied real estate on the planet"[8] – environmentalist lobbying caused the Obama administration and allied lawmakers to oppose the project, and in 2011 federal funding for it was revoked. The Government Accountability Office noted that no technical or safety reasons were provided for shutting down the project.[9] The Trump administration pledged to restart the project, but did not, and the Biden administration, while claiming that it sees climate change as an "existential crisis," (i.e. one that involves the *survival of humanity*) has chosen not to do so as well. Meanwhile, with funding revoked, the government faces a liability of $15 billion, growing by another billion every two years, for failing to meet its contractual obligations to produce a nuclear waste repository.[10]

## NUCLEAR ACCIDENTS

Nuclear accidents are certainly possible, but rare. Over the course of its entire history, the world's commercial nuclear industry has had three major accidents: one at Three Mile Island in Pennsylvania in 1979; one in Fukushima, Japan, in March 2011; and the other at Chernobyl in Ukraine in 1986.

The Three Mile Island event was the only nuclear disaster in US history. It is also unique in another sense; in that it was the only major disaster in world history in which not a single person was killed or even injured in any way.

There were two 843 MWe PWRs at three Mile Island, labeled TMI-1 and TMI-2. On March 28, 1979, the date of the accident, TMI-1 was shut down, but TMI-2 was operating at full power, when its turbine tripped. This shut off the secondary loop water flow to the steam generator, which meant that nothing was taking away heat from the primary loop cooling the reactor. As a result, the control rods dropped into place, shutting down the chain reaction instantly. But because the reactor had been operating for some time, a large inventory of highly radioactive fission products had built up in the core, and they continued to generate heat via radioactive decay at several percent the reactor rated power after shut down. So instead of the thermal power of the reactor dropping from 2500 MWt (the thermal power of a nuclear reactor is about three times its electrical power, because PWRs operate with an efficiency of 33%) to zero, it dropped instantly to 175 MWt, decreasing to 50 MWt an hour after shutdown, declining further to 20 MWt after 3 hours.

The fact that a reactor would continue to generate decay heat even after the chain reaction was shut down is and was well known. According to antinuclear activists, it meant that while loss of coolant would cause nuclear fission to cease, the uncooled reactor would melt itself down, with a mass of highly radioactive fission products unstoppably melting its way through the 20 cm (8") thick steel pressure vessel, then through the 2.6 meter (8.6 foot) thick containment building floor, then right on down through the Earth, all the way to China.

At TMI-2 this theory was put to the test, because while an emergency cooling system was in place to keep cooling water flowing into the reactor under such conditions, and it turned on automatically, the confused reactor operators turned it off. As a result, the reactor did melt down.

But instead of the hot fission products melting their way through the pressure vessel, the containment building, and the Earth, all the way to China, they actually melted their way a couple of centimeters (about an inch) into the pressure vessel and stopped there. That was it. A billion-dollar reactor was lost, but the containment system was never even seriously challenged. A few Curies of radioactive Iodine 131 gas (half-life 11 days) were vented, exposing the public in the area to about 1 mrem of radiation, equivalent to the extra dose they would have received spending a five-day ski trip to Colorado instead of staying in Pennsylvania. The environmental impact was zero. If anyone was harmed, it was because the very antinuclear lawyers running the Nuclear Regulatory Commission decided that the accident warranted keeping the untouched TMI-1 unit shut down for the next six years, and it is estimated that the pollution emissions over that time released by the coal-fired power plants used to replace its output were probably responsible for about 300 deaths.[11]

The 2011 Japanese accident was much more serious. Caused by a powerful undersea earthquake and resulting tsunami that buffeted the facility with waves nearly fifty feet high, the power plant flooded, and both grid power and the onsite backup diesel generators were knocked out, eliminating the emergency core cooling system. This eventually led to full meltdown of three of the six reactors. Nevertheless, if anything, the Fukushima event proved the safety of nuclear power. In the midst of a devastating disaster which killed some 28,000 people by drowning, falling buildings, fire, suffocation, exposure, disease, and many other causes, not a single person was killed by radiation. Nor was anyone outside the plant gate exposed to any significant radiological dose. There may have been substantial nuclear-related casualties, however—but these were caused by Gregory Jaczko, the chairman of the Obama administration's Nuclear Regulatory Commission. Following the Fukushima incident, Jaczko spread panic by warning all American citizens to stay at least fifty miles away from the Fukushima power plants. The resulting climate of fear and panic hampered rescue efforts. U.S. Navy forces steaming toward the rescue were ordered to stay 100 miles away –leaving unknown numbers of victims adrift on wreckage or trapped under buildings within the fifty-mile zone to die, despite a complete lack of evidence for any actually dangerous levels of radiation outside of the plant gate.[12]

(Jaczko may well be the most anti-nuclear chairman that the Nuclear Regulatory Commission has ever had. A former aide to the very anti-nuclear Representative Ed Markey (D.-MA) and to Senate Majority Leader Harry Reid (D-NV), Jaczko was responsible for the extraordinary efforts of the Obama administration to prevent the establishment of a permanent nuclear waste repository at Yucca Mountain, Nevada. In doing so, he not only broke his word to Congress—which, concerned about bias stemming from his work for Harry Reid, asked him at his confirmation hearing to recuse himself from Yucca Mountain matters—but violated the fundamental purpose of his office. Instead of trying to make nuclear power as safe as possible, Jaczko's effort to stop safe nuclear waste storage located far away from populated areas represented an attempt to make the industry as dangerous as possible, in order to shut it down.)[13]

From the point of view of radiation release, Chernobyl was the most serious nuclear-plant disaster of all time. At Chernobyl, the reactor actually had a runaway chain reaction and disassembled, breaching all containment. Approximately fifty people were killed during the event itself and the fire-fighting efforts that followed immediately thereafter. Furthermore, radioactive material comparable to that produced by an atomic bomb was released into the environment. According to a study by the International Atomic Energy Agency and World Health Organization using LNT methodology, over time this fallout could theoretically cause up to four thousand deaths among the surrounding population. Chernobyl was really about as bad as a nuclear accident can be. Yet, even if we accept the grossly exaggerated casualties predicted by LNT theory as being correct, in comparison to all the deaths caused *every year* as a result of the pollution emitted from coal-fired power plants, its impact was minor. Chernobyl-like catastrophes would have to occur *every day* to approach the toll on humanity currently inflicted by coal. By replacing a substantial fraction of the electricity that would otherwise have to be generated by fossil fuels, the nuclear industry has actually saved countless lives.[14]

Still, Chernobyl events need to be prevented, and they can be, by proper reactor engineering. In the first place, the Chernobyl reactor had no containment building. If it had, there would have been no radiological release into the environment. In the second place, had the reactor been designed to lose reactivity beyond its design temperature – as all water-moderated reactors are – the runaway reaction would never have occurred at all. The key is to design the reactor in such a way that as its temperature increases, its power level will go *down*. In technical parlance, this is known as having a "negative temperature coefficient of reactivity." Water is necessary for a sustained nuclear reaction because it serves to slow down, or "moderate," the fast neutrons born of fission events enough for them to interact with surrounding nuclei to continue the reaction. (Just like an asteroid passing by the Earth, a neutron is more likely to be pulled in to collide with a nucleus if it is going slow than if it is going fast.) It is physically impossible for such a water-moderated reactor to have a runaway chain reaction, because as soon as the reactor heats beyond a certain point, the water starts to boil. This reduces the water's effectiveness as a moderator, and without moderation, fewer and fewer neutrons strike their target, causing the reactor's power level to drop. The system is thus intrinsically stable, and there is no way to make it unstable. No matter how incompetent, crazy, or malicious the operators of a water-moderated reactor might be, they can't make it go Chernobyl.

In contrast, the reactor that exploded at Chernobyl—a Soviet RBMK reactor—was moderated not by water, but by graphite, which does not boil. It therefore did not have the strong negative temperature reactivity feedback of a water-moderated system, and in fact, because water absorbs neutrons while graphite does not, it actually had a *positive* temperature coefficient of reactivity, which caused power to soar once water coolant was lost. It was thus an unstable system, vulnerable to a runaway reaction when its operators decided to do some really dumb experiments. Furthermore, with a huge amount of hot graphite freely exposed to the environment once the reactor was breeched, fuel was available for a giant bonfire to send the whole accumulated stockpile of radioactive fission products right up into the sky. The Chernobyl reactor wasn't just unstable, it was flammable!

No such crazy system could ever get permitted in the United States or any other civilized country by the local fire department, let alone the nuclear regulators. Those who died at Chernobyl weren't victims of nuclear power. They were victims of the Soviet Union.

## CAN REACTORS EXPLODE LIKE BOMBS?

Chernobyl was a runaway fission reaction, but it was not an atomic bomb. The strength of the explosion was enough to blow the roof off the building and break the reactor apart into burning graphite fragments, but the total explosive yield was less than that provided by a medium-sized conventional bomb. But of course, the reactor operators weren't *trying to* achieve a Hiroshima. They were just goofing around. What could they have done had they been *trying*?

I know this question is very much on the minds of those of my readers who are terrorists, and who would like to explore the possibilities of detonating reactors as bombs, particularly in the United States. So let's see what would be involved in trying to make an American PWR explode like a nuclear weapon.

The first thing you are going to have to do is take over the plant. They are heavily defended, so you are going to need at least a company of well-trained mercenaries, complete with the full range of modern infantry weapons, ammunition, rations, personal protective equipment, and MRAPs for transport. These are available from the Blackwater Corporation, and as their highly capable personnel are innocent of any complicating loyalties or ethical codes, they will happily do the job for you provided you can pay their fair market price. But be sure not to let them know that you intend to turn the plant into a nuclear fireball once they are inside, or they may charge more. (Wagner Group mercenaries are cheaper, but they are under international sanctions. Employing them to seize and explode a nuclear power plant could potentially embroil you in serious legal complications. Furthermore, many of their men are of low quality, having gotten AIDS or other exotic diseases from each other during their pre-recruitment prison terms.)

So, assuming that you have chosen qualified mercenary assistance and your assault is successful, now you are inside and in full control of the facility. What can you do? You could pull all the control rods out of the reactor, which will ramp up the chain reaction. But as soon as the water boiling becomes too vigorous, it will no longer function adequately as a moderator, and the chain reaction will stop growing, leaving you with a constant power reactor. You could drain the water out of the reactor, but then the chain reaction will stop altogether, and all you will be left with is decay heat. In that case, the reactor will melt down, just as it did at Three Mile Island, leaving you with some melted stuff sitting at the bottom of the pressure vessel, as occurred there. Big whoop. If you can get your mercenaries to blast apart in the four-foot-thick reinforced concrete containment building and the 8-inch-thick steel pressure vessel without asking any questions, you might manage to release some radiological material to the environment. But since the reactor is not made of flammable graphite, you won't be able to set a big fire to spread its contents around fast enough to do much harm before the authorities or the news media figure out that something is going on and warn the people downwind of you to drive away. How are you going to explain such a pathetic return for your effort to your financiers?

Basically you have a problem. A bomb explosion needs to be done using fast neutrons. Slow neutrons take much too long to multiply, because each generation must go through dozens of collisions to bring them down to thermal energies where the uranium nuclei have big fission cross sections. But once the chain reaction power has reached a level where the system starts to disassemble, time is something you don't have. As soon as the bomb breaks apart, the chain reaction will stop. So the best you can do with slow reactions is a Chernobyl-like fizzle. To make a bomb, you need to use fast neutrons, because only they can multiply fast enough. But – as you can see from the data in Table 5.2 – while uranium and plutonium nuclei have slow neutron fission cross sections of 500 barns or more, their fast neutron cross sections are just a barn or two. So the only way to get a fast fission chain reaction happen at all is to have material that is highly enriched – ideally 90 percent U-235 or better. But PWR fuel is only 3 percent enriched, so it just won't work. From your point of view as a bombmaker, the stuff is junk. You can no more make it explode like a nuclear bomb than you can a bag of apples.

Furthermore, even if you could get your hands on a fast reactor using highly enriched U-235, for example by stealing it out of a nuclear submarine while no one is looking, you still won't be able to explode it as a bomb. Exquisite engineering design is required to bring a critical mass of highly enriched fissile materials together so fast that heat generated by a partial chain reaction cannot blow them apart before they can combine, while having the chain reaction be fast enough so that it will all go off before the materials can separate. In fact, it took the concerted efforts of some of the world's greatest scientists working at Los Alamos to design and implement such a controlled "implosion" system during World War II. A reactor has no such mechanism.

So, I'm sorry to be so negative, but it just can't be done.

## NUCLEAR PROLIFERATION

Well, you say, asking for a friend, couldn't the industrial infrastructure used to produce three percent enriched fuel for nuclear reactors also be used to make 90 percent enriched material for bombs?

Yes, it's true. Natural uranium contains 0.7 percent uranium-235 ($^{235}$U), which is capable of fission, and 99.3 percent uranium-238 ($^{238}$U), which is not. In order to be useful in a commercial nuclear reactor, the uranium is typically enriched to a 3 percent concentration of $^{235}$U. The same enrichment facilities could indeed be used, with some difficulty, to further concentrate the uranium to 93 percent $^{235}$U, which would make it bomb-grade. Additionally, once the controlled reaction begins, some of the $^{238}$U will absorb neutrons, transforming it into plutonium-239 ($^{239}$Pu), which is fissile. Such plutonium can be reprocessed out of the spent fuel and mixed with natural uranium to turn it into reactor-grade material. Under some circumstances, it could also be used to make bombs instead. Thus, the technical infrastructure required to support an end-to-end nuclear industry fuel cycle could also be used to make weapons.

It is also true, however, that such facilities could be used to make bomb-grade material without supporting any nuclear reactors. In fact, until Eisenhower's Atoms for Peace policy was set forth, the AEC opposed nuclear reactors precisely because *they represented a diversion of fissionable material from bomb making.* If plutonium is desired, much better material for weapons purposes can be made in standalone atomic piles than can be made in commercial power stations. This is so because when Pu-239 is left in a reactor too long, it can absorb a neutron and become Pu-240 which is not fissile, and extremely difficult to separate from P-239. Furthermore while not fissile from reaction with neutrons, Pu-240 undergoes spontaneous fissions as a form of decay. Inserted in bomb material, it could set the bomb off prematurely. Commercial power operators don't want to constantly be shutting down in order to remove lightly used fuel from their reactors. In consequence the Pu-240 content of used civilian reactor fuel builds up to about 26% of the Pu-239. But if it's more than 7% it makes the plutonium useless in a bomb (and in fact the US military spec requires it to be less than 1%.) Both the United States and the Soviet Union had thousands of atomic weapons before either had a single nuclear power plant, using either highly enriched uranium or plutonium made in special military fuel production reactors that allow constant removal of fuel. Others desirous of obtaining atomic bombs could and would proceed the same way today.[15]

As an additional safeguard against nuclear proliferation, thorium reactors can be used in place of uranium reactors. Thorium (atomic number 90, atomic weight 232) is about four times as plentiful as uranium, while being about 1/3 as radioactive. It can endure very high temperatures, which is why it was commonly used as the mantle in kerosene lanterns. It is not fissile, but fertile. That is, when placed in a nuclear reactor it absorbs neutrons, transforming itself into uranium-233 in the process. Uranium-233 makes excellent reactor fuel, but because some of it decays by neutron emission into uranium-232, it can set off a bomb prematurely. Furthermore, uranium 232 is an emitter of hard gamma rays, making a weapon containing it unacceptable in a military environment (and easily detectable in a terrorist device.) These two features make U-233 bred in thorium reactors useless for bombs. In fact, the uselessness of U-233 for bomb-making is exactly why the military-dominated early nuclear programs in the US and Soviet Union avoided thorium reactors. However, in the late 1970s, the Carter administration became interested in promoting proliferation-proof reactors, and commissioned Admiral Rickover to take the original Shippingport reactor and convert it to U-233/thorium. This demonstration was entirely successful. However, with the halt in new reactor orders in the USA following the Three Mile Island event, it has been left to India, which is uranium short but very thorium rich, to move forward with the technology.

I like to kid around in this book, because I believe that reading should be fun. But I take the matter of weapons of mass destruction very seriously, and I want to say something very serious about it to you, dear reader, right now.

Due to advances in biology, it is now possible to make biological weapons that are far more dangerous than nuclear devices, at many orders of magnitude lower cost. During the recent CV19 epidemic, there were allegations that the virus was made by altering a native strain found in bats in the biological research facility in Wuhan, China.[16] These claims are unproven and sharply disputed. What is not disputed, however, is that it *could* have been made there. Think about that: a facility costing less than 1/1000th of the infrastructure needed to produce a critical mass of bomb grade material can be used to create a virus that kills millions.

That is the reality of the modern age. Scientific knowledge has given us extraordinary powers of creation, and therefore destruction as well, and there is no way to unknow it.

If we wish to avoid catastrophe, we need to build a world which offers plenty for everyone.

That's why we need nuclear power.

# CHAPTER 9

# HOW TO CUT COSTS

*"The first thing we do is kill all the lawyers."*
— William Shakespeare, Henry VI, Part 2.

**IF YOU HAVE** built your nuke as I instructed, you should now be generating power at a small fraction of the cost of fossil fuel providers. Your 1000 MWe nuke probably cost something like $500 million, which is about what it takes to build a few suburban shopping malls. Yet the rest of the industry is shelling out $5 billion or more — ten times as much to build a plant of the same size? What gives? Why do commercial nukes now cost so much? According to antinuclear analysts, that's just the way it has to be. So, regrettably, unfortunately, the dream of cheap unlimited non-polluting power is just a fantasy, and the little people will simply have to learn how to get by with less.

Bunk. New nuclear power plants do indeed cost a lot, but they shouldn't, and they didn't use to. The *Nautilus* only took four years to develop and build, and the Shippingport plant three. In the 1960s, power stations got much larger, but it still only took an average of five years to build a new nuclear power in the United States. In South Korea today it still takes an average of four years to build a nuclear power plant. Now, as a result of environmentalist efforts to introduce regulatory roadblocks at every stage of the construction process, it can take 16 years to build one in the West. This has enormously increased the cost of all such projects.

The transition from nuclear power plants taking four years and hundreds of million to build to plants taking 16 years and many billions to build took place during the 1970s and 1980s. In the early 1990s, Bernard Cohen, a professor of nuclear physics at the University of Pittsburgh took a look at what was driving up the costs.[1] He wrote:

"Several large nuclear power plants were completed in the early 1970s at a typical cost of $170 million, whereas plants of the same size completed in 1983 cost an average of $1.7 billion, a 10-fold increase. Some plants completed in the late 1980s have cost as much as $5 billion, 30 times what they cost 15 years earlier. Inflation, of course, has played a role, but the consumer price index increased only by a factor of 2.2 between 1973 and 1983, and by just 18% from 1983 to 1988. What caused the remaining large increase? Ask the opponents of nuclear power and they ... create the impression that people who build nuclear plants are a bunch of bungling incompetents. The only thing they won't explain is how these same 'bungling incompetents' managed to build nuclear power plants so efficiently, so rapidly, and so inexpensively in the early 1970s."

Cohen goes on to list many examples, which I have laid out in Table 9.1. The massive increase in nuclear power plant costs following the imposition of an antinuclear regulatory structure by the 1977-1980 Carter administration is instantly observable.

## TABLE 9.1 RISE IN COST OF NUCLEAR POWER PLANT CONSTRUCTION

| Utility | Plants | Year Completed | Cost per Kilowatt |
|---|---|---|---|
| Commonwealth Edison | Dresden | 1970-71 | $146/kW |
| Commonwealth Edison | Quad Cities | 1973 | $146 |
| Commonwealth Edison | Zion | 1973-74 | $280 |
| Commonwealth Edison | LaSalle | 1982-84 | $1,160 |
| Commonwealth Edison | Byron | 1985 | $1,880 |
| Commonwealth Edison | Braidwood | 1987 | $1,880 |
| Northeast Utilities | Milestone 1 | 1971 | $153 |
| Northeast Utilities | Milestone 2 | 1975 | $487 |
| Northeast Utilities | Milestone 3 | 1986 | $3,326 |
| Duke Power | Oconee | 1973-74 | $181 |
| Duke Power | McGuire | 1981-84 | $848 |
| Duke Power | Catauba | 1985-87 | $1,703 |
| Philadelphia Electric | Peach Bottom | 1974 | $292 |
| Philadelphia electric | Limerick | 1988 | $2,560 |

"A long list of such price escalations could be quoted, and there are no exceptions," Cohen concludes. "Clearly, something other than incompetence is involved."

If one examines Cohen's data, it becomes crystal clear that the central cause of the price increase has been a radical increase in construction time. This can be seen in Table 9.2

**TABLE 9.2 INCREASING CONSTRUCTION TIME MULTIPLIED NUCLEAR POWER PLANT COSTS**

| Year Begun | Time | Cost/kW current$ | Cost/kW 2021$ | $2021/time squared |
|---|---|---|---|---|
| 1967 | 5.5 | 165 | 1292 | 42.7 |
| 1968 | 5.75 | 219 | 1645 | 49.8 |
| 1969 | 6 | 280 | 1996 | 55.6 |
| 1970 | 6.5 | 326 | 2199 | 52.0 |
| 1971 | 7 | 375 | 2423 | 49.4 |
| 1972 | 7.5 | 473 | 2958 | 52.6 |
| 1973 | 8 | 570 | 3357 | 52.5 |
| 1974 | 8.5 | 720 | 3823 | 52.9 |
| 1975 | 10 | 882 | 4287 | 42.9 |
| 1976 | 10 | 964 | 4432 | 44.3 |
| 1977 | 10 | 1000 | 4320 | 43.2 |
| 1978 | 10 | 1092 | 4379 | 43.8 |
| 1979 | 11 | 1755 | 6318 | 52.2 |
| 1980 | 12 | 2706 | 8578 | 59.6 |

The first three columns in Table 9.2 show data published by Cohen. In this third column, I translate his current-year dollars into inflation adjusted 2021 dollars. In the last column, I show what you get if you divide the inflation adjusted cost of the power plants by the square of the number of years it took to complete them. As you can see, over a period of thirteen years during which the cost of nuclear power plants, measured in current dollars, increased seventeen-fold, this parameter remained virtually constant. In other words, construction time grew, and as it did so, *the cost of the plants, in inflation-adjusted dollars,* **rose as the square of the time!**

There are many reasons why stretching out time increases costs. In the first place, it multiplies labor costs, because workers standing around waiting for orders to do the next task cost just as much as those actually doing work. Stretching out time also adds to interest rate costs, and inflation costs. In addition, however, nuclear power plant construction time delays are frequently caused by design changes ordered by regulators, or litigation launched by anti-nuclear intervenors, both of which can be extremely costly themselves

The regulatory whipsawing and strangulation of the nuclear industry has not served to advance safety. On the contrary, it has made it next to impossible to correct problems or introduce improvements. I saw this at first hand in the late 1980s when I was working for the Washington state Office of Radiation Protection. Because it was just over the state line in Oregon, we participated in regulatory oversight on the Trojan nuclear power plant. This excellent plant had been built quickly in the 1970s, and so was able to produce electricity at the very attractive cost of 2 cents per kilowatt hour – much less than fossil fuels and fully competitive with the cheap hydroelectric power available in the Pacific Northwest. But it had a problem with corrosion of the pipes in its secondary loop. This subsystem transformed nuclear thermal power, taken outside the reactor by the primary loop, into steam to drive the turbine generators. This was forcing the utility to shut down for several months every year to replace the corroded pipes.

The utility's management understood the source of the problem and wanted to solve it be replacing the carbon steel pipes in the secondary loop with stainless steel, which would not corrode. We and our counterparts in the Oregon state government supported them in this decision, as it was obviously the right thing to do. But the federal Nuclear Regulatory Commission told them they could not do so, because their operating license specified carbon steel for those pipes, not stainless steel. If they wanted to replace the carbon steel with stainless, they would have to apply for a new operating license, a process which by the 1980s had been made so laborious and perilous with opportunities for hostile legal interventions as to be out of the question. So, prevented by federal regulators from correcting its problem, the plant lumbered on, until it was finally shut down for good in 1992.

The overall process used by the Nuclear Regulatory Commission to strangle the nuclear industry goes by the charming term of "ratcheting." As Cohen said, "Like a ratchet wrench which is moved back and forth but always tightens and never loosens a bolt, the regulatory requirements were constantly tightened, requiring additional equipment and construction labor and materials. ...between the early and late 1970s, regulatory requirements increased the quantity of steel needed in a power plant of equivalent electrical output by 41%, the amount of concrete by 27%, the lineal footage of piping by 50%, and the length of electrical cable by 36%. The NRC did not withdraw requirements made in the early days on the basis of minimal experience when later experience demonstrated that they were unnecessarily stringent. Regulations were only tightened, never loosened. The ratcheting policy was consistently followed...

Cohen then goes on to list examples of how regulatory ratcheting increased both the time from project initiation to ground breaking and from ground breaking to operation testing. I present his data in Table 9.3.

## TABLE 9.3 GROWTH OF DELAYS IN NUCLEAR POWER PLANT CONSTRUCTION

| Project Phase | Year | Time Required |
|---|---|---|
| Project Initiation to Ground Breaking | 1967 | 16 months |
| Project Initiation to Ground Breaking | 1972 | 32 months |
| Project Initiation to Ground Breaking | 1980 | 54 months |
| Groundbreaking to Operation Testing | 1967 | 42 months |
| Groundbreaking to Operation Testing | 1972 | 54 months |
| Groundbreaking to Operation Testing | 1980 | 70 months |

As a result of these delays, regulatory ratcheting *quadrupled* the inflation-adjusted cost of nuclear power construction from the sixties to the eighties.

This massive addition of regulatory overhead did not help advance nuclear technology. Quite the contrary, says Cohen, if anything, it made things worse. "What has all this bought in the way of safety? One point of view often expressed privately by those involved in design and construction is that it has bought *nothing*. A nuclear power plant is a very complex system and adding to its complexity involves a risk in its own right. If there are more pipes, there are more ways to have pipe breaks, which are one of the most dangerous failures in reactors. With more complexity in electrical wiring, the chance for a short circuit or for an error in hook-ups increases, and there is less chance for such an error to be discovered. ... It is difficult to determine whether or not reducing a particular safety problem improves safety more than the added complexity reduces safety."

In fact, there was no justification for the post 1973 regulatory onslaught. The nuclear power plants of the 1960s worked fine, with no accidents, and the regulators of that time were quite satisfied with them. There were some justified regulation changes based on lessons learned from the Three Mile Island accident, but those only added a few percent to the cost of plants. There were no scientific findings or technical developments that called for changing the regulations. In fact exactly the opposite was true, since it was found that the emergency core cooling systems of nuclear power plants worked better than the baseline requirements, and far better than the worst-case estimates advanced by antinuclear activists. The whole ratcheting campaign was arbitrary, and contrary to any normal regulatory process, it only worked one way, with no provision to eliminate regulations that were found to be unnecessary. As Cohen put it:

"The ratcheting effect, only making changes in one direction, was an abnormal aspect of regulatory practice unjustified from a scientific point of view. It was a strictly *political* phenomenon that quadrupled the cost of nuclear power plants, and thereby caused no new plants to be ordered and dozens of partially constructed plants to be abandoned."

Even worse than the NRC's constant tightening of regulations was the fact it was always changing them. As regulations changed, designs needed to be changed, forcing construction work to be altered or reversed in the middle of the job. There is no more effective way to cause a catastrophic increase in the cost of a construction project than to change its design after the work in underway. As Cohen explained:

"As anyone who has tried to make major alterations in the design of his house while it was under construction can testify, making these changes is a very time-consuming and expensive practice, much more expensive than if they had been incorporated in the original design. In nuclear power plant construction, there were situations where the walls of a building were already in place when new regulations appeared requiring substantial amounts of new equipment to be included inside them. In some cases this proved to be nearly impossible, and in most cases, it required a great deal of extra expense for engineering and repositioning of equipment, piping, and cables that had already been installed. In some cases it even required chipping out concrete that had already been poured, which is an extremely expensive proposition.

"Constructors, in attempting to avoid such situations, often included features that were not required in an effort to anticipate rule changes that never materialized. This also added to the cost. There has always been a time-honored tradition in the construction industry of on-the-spot innovation to solve unanticipated problems; the object is to get things done. The supercharged regulatory environment squelched this completely, seriously hurting the morale of construction crews. For example, in the course of many design changes, miscalculations might cause two pipes to interfere with one another, or a pipe might interfere with a valve. Normally a construction supervisor would move the pipe or valve a few inches, but that became a serious rule violation. He now had to check with the engineering group at the home office, and they must feed the change into their computer programs for analyzing vibrations and resistance to earthquakes. It might take many hours for approval, and in the meanwhile, pipefitters and welders had to stand around with nothing to do.

"Requiring elaborate inspections and quality control checks on every operation frequently held up progress. If an inspector needed extra time on one job, he was delayed in getting to another. Again, craft labor was forced to stand around waiting. In such situations, it sometimes pays to hire extra inspectors, who then have nothing to do most of the time... Cynicism became rampant, and morale sagged"

"Changing plans in the course of construction is a confusing process that can easily lead to costly mistakes. The Diablo Canyon plant in California was ready for operation when such a mistake was discovered, necessitating many months of delay.

Delaying completion of a plant typically costs more than a million dollars per day."

Think about that, through arbitrary actions, the NRC repeatedly caused multi-year delays. At a million dollars per day, a three-year delay will inflict over a billion bucks worth of damage to a project. Yet the NRC faced no accountability for such massive acts of vandalism against the industry.

The environmentalist-infested NRC did further extraordinary harm to nuclear power by arranging its regulatory structure so that a nuclear plant would first get a construction license, and then after construction was underway, open up hearings to provide opportunities for activists to litigate against permits for further work at every stage of the process, with the plant's owners only able to apply for an operating license after it was complete.

Anti-nuclear groups took full advantage of the excellent opportunities their cohorts within the bureaucracy had provided them, and imposed years of costly delays on projects coast to coast through endless hearings and court actions.

The delays also multiplied cost escalation by creating cash flow problems for utilities. Utilities only are able to set aside so much cash to do a project, and they plan it accordingly. But if delays cause costs to rise, a utility can be forced to suspend construction for a time, which can send the final cost of the project through the roof.

The effect of this campaign of regulatory strangulation can be seen in Fig. 9.1, adapted from information presented in a very data-rich 2016 paper by the Breakthrough Institute's Jessica Lovering, Arthur Yip, and Ted Nordhaus, which shows the exponential takeoff of the cost of US nuclear power plants starting in the late 1970s.[3] It can be seen that costs multiplied tenfold from $700/kWe for plants started in 1968 to $7000/kWe for those begun in the late 1970s. There was no technical reason for this jump, its causes were purely political.

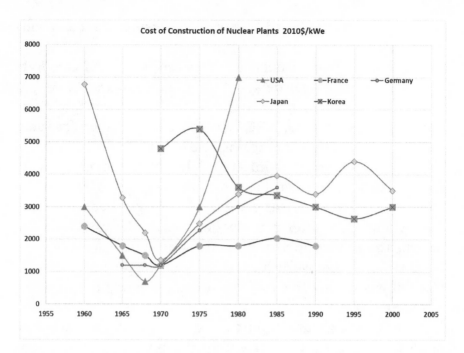

**Fig. 9.1** As a result of hyper regulation, costs of US nuclear power plants took off exponentially in the 1970s. Simultaneously French costs remained stable while Korean costs declined. The dates given are the dates of *start* of construction. The cost lines, in units of $/kWe, are the average of numerous points presented individually in multiple scatter charts in reference 3, using current exchange rates and 2010 inflation adjusted dollars

In contrast, in France, where intervenor lawsuits are not allowed after construction begins, nuclear power plant costs remained stable. In South Korea costs declined, as would be expected in a new industry gaining experience.[4]

But the worst sabotage has been done by politicians playing to the anti-nuclear crowd. For example, after some $6 billion had been spent to build and complete the Shoreham nuclear power plant on Long Island in 1984, New York governor Mario Cuomo (father to the more recent disaster of the same last name) capriciously ordered that it could not be opened, turning the project into a total economic loss. (Actually he did offer to buy it for $1.)[5] The threat of similar wrecking operations by political malefactors elsewhere has made financing nuclear power plant construction extremely difficult.

Thus the case made by antinuclear activists that nuclear power is too costly is as sincere as that of the boy, who, convicted of murdering his parents, asked the judge to let him off on the grounds that he was an orphan.

If humanity is to have an open future, these people need to be exposed for the frauds they are.

## FOCUS SECTION: MERCENARY ENVIRONMENTALISM

There are three principal reasons why the leaders of putative environmentalist movement have been committed to destroying the nuclear industry for the past half century.

One is because they are true adherents or antihuman ideology, believing that humanity is fundamentally a race of vermin, whose number, activities, and liberties must be severely constrained in order to protect a statically conceived "Nature."[6] Liberating humanity with unlimited energy would upend that grand project.

Another, more immediate concern, is that the environmentalist organizations require pollution for agitational purposes. Accordingly, they hate nuclear energy because it would solve a problem they need to have.

The third motivation, however, is the most compelling of all. They work to destroy the nuclear industry because they get paid to do so.

Let's face it. Business is business. There is nothing as sincere as cash. Anyone who comes along offering to sell electricity at lower than the going rate is going to sincerely upset a lot of people. The world electricity market is worth trillions. Its owners will do whatever it takes to defend their interests.

They have therefore not hesitated to hire the services offered by the environmentalist movement.

The primary environmentalist organizations are major businesses themselves. These include the Sierra Club, Greenpeace, the Natural Resources Defense Council, the Environmental Defense Fund, Friends of the Earth, the Nature Conservancy, and the World Wildlife Fund. The business services some offer include conducting campaigns that ruin their donors' competitors.

The income and assets of some of these organizations of presented in Table 9.2 below.

## TABLE 9.2 THE INCOME AND ASSETS OF SOME MAJOR ENVIRONMENTAL ORGANIZATIONS[7]

|  | Income | Assets |
|---|---|---|
| Sierra Club | $191 million | $232 million |
| Friends of the Earth | $10 million | $15 million |
| Greenpeace | $54 million | $26 million |
| Natural Resources Defense Council | $182 million | $411 million |
| Environmental Defense Fund | $211 million | $245 million |
| World Wildlife Fund | $257 million | $503 million |
| Nature Conservancy | $1,184 million | $7,410 million |

In characterizing these organizations as businesses, I do not wish to discount their ideological motivations. They are sincerely committed to ending the progress of human civilization, or at least denying its benefits, such as electricity, opportunity, proper nutrition, and good health, to those parts of the world who do not yet have them. Thus, they have waged campaigns to deny life-saving pesticides, blindness-preventing vitamin-enriched foods, the right to use land, and the right to have children, to Third World people on the basis of purely aesthetic considerations.[8]

But it is possible to serve Satan and Mammon at the same time. Having assembled powerful mobs well-disposed towards destroying industries, the most financially astute environmentalist leaders have very profitably put their unique assets out for hire.

While it is true that from a purely ideological point of view, the environmentalists might wish to destroy all industries, from a practical standpoint this is impossible. In the real world, it is necessary to choose one's battles, and bills need to be paid. The environmentalist leaders are very sharp people, with a good

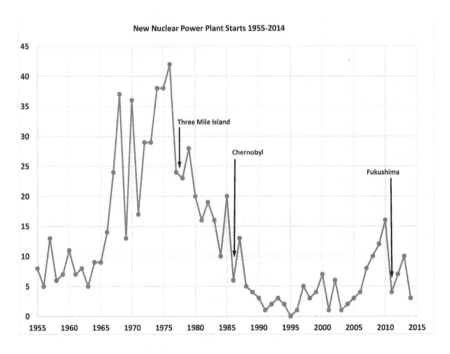

**Figure 9.2** Global nuclear reactor construction starts, 1955 to 2014. Note that the industry collapse began *before* Three Mile Island, and despite the 1974 oil price rise that made it *more* competitive. (Adapted from IAEA data[9])

understanding of these facts of life. Accordingly, their target selection process is donor driven.

Up through the 1960s, the primary environmentalist organizations were not particularly antinuclear, as the infant industry was unimportant. But in the early 1970s, two factors changed this situation. One was the advent of practical large nuclear power plants with very low costs. Plants begun in 1968 and completed in the early 1970s had construction costs of about $800/kWe (in 2020 dollars), allowing them to provide electricity at rates around 2 cents/kWh. This was less than fossil fuels and fully competitive even with cheap hydroelectric power. As a result, nuclear power reactor orders soared, with construction of 315 units begun worldwide between 1966 and 1975. (See figure 9.2)

Then, as a result of the 1973-74 Arab Oil Embargo, the price of oil quadrupled. The oil companies made a bundle off of this. But while oil's role as a transportation fuel was unassailable, its position as a resource for electricity was endangered by nuclear power. The inexpensive nuclear power plants coming into operation in the early to mid-70s threatened coal interests (who did not have a transportation fuel fortress) as well. But with plenty of cash on hand to pay for armies of environmentalist activists available for direction by the highest donor, the fossil fuel industry was able to fight back.

The way for the fossil fuel support for antinuclear activism had been prepared in 1968, when Sierra Club Executive Director David Brower, viewing the new technology from the point of view of ideological antihumanism, broke with organization over its preference for nuclear over coal to found the antinuclear Friends of the Earth. When, a few years later, he landed a substantial grant (about $500,000 in today's money) from the Atlantic Richfield oil company (aka Arco), the Sierra Club must have realized it was missing out on a good thing. It switched as well.

In 1974, the Sierra Club formally reversed its position on nuclear power, justifying its opposition to the non-polluting alternative to fossil fuels on the grounds that the use of inexpensive nuclear power could lead to "unnecessary economic growth." The Club soon laid out the strategy for its campaign. It would seek to cripple the industry by preventing the establishment of a nuclear waste disposal site, and inflate its costs by using a disinformation panic campaign to inspire industry-strangling hyper regulation. As Sierra Club Executive Director Michael McCloskey wrote at the time in a memo to the board, "Our campaign stressing the hazards of nuclear power will supply a rationale for increasing regulation... and add to the cost of the industry."[10]

This policy change proved to be a wise choice, from a fiduciary point of view, as the Club was rewarded for its wisdom with a $1 million donation from the Exxon corporation (~$5 million today) shortly afterwards. The rest of the environmental movement did not wait long to make their own moves to get a piece of the action.

The effect of this campaign can be seen in Figure 9.2. What is particularly noteworthy here is that the collapse of the nuclear industry occurred *before* the Three Mile Island accident. Rather it began shortly after the oil price rise that actually made it *more* competitive. This was because of the regulatory blitzkrieg unleashed by the antinuclear campaign.

While the antinuclear campaign stunted the growth of the nuclear power industry, it did not succeed in saving oil's position within power generation. In the 1973, oil produced 20 percent of American electricity while nuclear power produced 3 percent. Today those numbers are precisely reversed, with nuclear generating 20 percent of electricity to oil's 3 percent. However the position of coal as the largest supplier of US electric power was preserved with benign support from the 1970s antinuclear crusaders, who termed it a "bridge fuel" to the future solar powered utopia.[11]

This basic arrangement has continued to the present day, with significant changes including the switch of the favored bridge fuel from coal to natural gas, and the addition of green energy huckster funds to support the anti-nuclear crusade. Coal remained the dominant US electric power supplier until quite recently, supporting 40 percent of the American grid in 2012. Coal has since been added to the target list of the US environmentalists. It is still supported in preference to nuclear by most European Greens, despite their loudly voiced concerns for the "existential threat" posed by global warming. Indeed, as a result of the loony insistence of its Green Party that Germany's nuclear power stations be replaced by coal, "green" Germany currently produces six times as much carbon emissions per unit electricity generated as does 75 percent nuclear-powered France.

Biomass, by far the most polluting and environmentally destructive power source, however, is still supported by both American and European greens. American environmentalists have also turned against the US oil industry, preferring that the nation's fuel needs be met by Middle eastern suppliers instead. This latter turn has been reflected in the policy of the Biden administration, which while seeking to shut down US oil production has called upon OPEC to increase its output. Apparently Saudi oil is carbon free.

As Michael Shellenberger has documented at length is his excellent book, *Apocalypse Never*,[12] the funds being transferred to support such remarkable policies are quite substantial. Nine figure donors include natural gas investor Michael Bloomberg, gas and coal magnate Tom Steyer, Chesapeake Energy chief Aubrey McClendon, and the government of Qatar, which alongside its support of the ISIS terrorist organization, generously gifted global environmental savoir Al Gore a tidy sum of $100 million.

The receivers of this beneficence have also gotten in on the action themselves, with noteworthy antinuclear campaigners the Natural Resource Defense Council and the Environmental

Defense Fund holding natural gas portfolios worth hundreds of millions of dollars themselves.

It would be tedious here to go through all the dirty deals and transactions involved in securing environmentalist support for replacing nuclear energy with fossil fuel power sources. For that, I refer you to Shellenberger.

However, if you want to understand how it is possible that a group of organizations who claim that the *existence of the human race is threatened* (that's what "existential crisis" means) by carbon emissions should so fanatically oppose their antidote, I suggest you employ what experience has proven to be the best possible method of criminal investigation.

Follow the money.

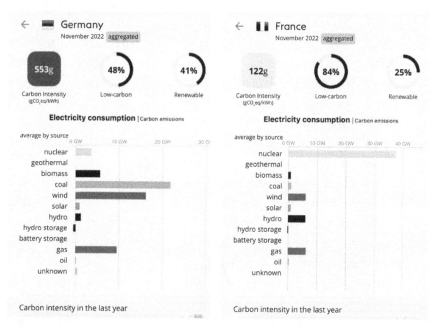

**Fig. 9.3** Comparison of German and French average electric power $CO_2$ emissions for the period December 2021 through November 2022. "Green" Germany has five times the emissions per unit electricity as nuclear France. (Credit: Electricity Maps)

# CHAPTER 10

# BREEDING MORE FUEL THAN YOU BURN

EVEN WHILE THE World War II was still underway, scientists working on the Manhattan Project began to think about how the titanic new source of power they were unleashing to serve as a means of destruction could be used for creation as well. Nuclear power offered humanity an energy source millions of times more potent than that of fossil fuels. Imagine if the only application anyone had ever found for gasoline was the manufacture of napalm. Limiting nuclear power to bombs would be a historical tragedy even more catastrophic.

But if nuclear power was to substantially supplant fossil fuels as the energy foundation for industrial society, let alone lead to a greater world beyond, the supply of nuclear fuel would have to be at least comparable to that of oil and coal reserves. How much uranium actually existed?

In 1944, Manhattan Project scientist Phil Morrison conducted a study and came to a pessimistic conclusion. According to the Morrison study, the world possessed, at most, a few thousand tons of useful uranium ore.[1] This estimate was wildly incorrect. Current estimates for world uranium ore reserves are 5.3 million tons, with many billions of tons more available dissolved in sea water or present on land in granite and other minerals. While the uranium concentration in sea water and rocks is low, the energy yield it provides is orders of magnitude greater than that required for its extraction.

Uranium seemed rare in the 1940s, because up until that time no one had bothered looking for it. It was not until the 1960s, by which time a fair amount of looking had been done, that the nonexistence of the putative uranium shortage became evident. In the meantime, scientists hoping to make nuclear power a source of abundant energy to power civilization thought that uranium scarcity was a problem they needed to solve. So they did.

Natural uranium consists of 0.7% U-235 and 99.3% U-238. Only the U-235 is readily fissile. In making the Hiroshima bomb, which was enriched to 80% U-235 to enable it to perform a fast runaway chain reaction, the Manhattan Project had thrown away more than 100 kilograms of U-238 for every kilogram of U-235 it had kept for use in the bomb.

But while not readily fissile, U-238 is fertile. After it absorbs a neutron it becomes plutonium-239, which is about as fissile as U-238. Being a different element than uranium, such plutonium can be readily separated from the surrounding U-238. This is how the Nagasaki bomb was made.

Depending on conditions, each nuclear fission reaction produces between two and three neutrons. Let's say the average in a certain reactor employing 3% enriched U-235 in 97% U-238 is 2.1 neutrons per fission. To maintain a critical chain reaction, one of these must be used to set off another fission. That leaves 1.1 neutrons that are not needed to sustain power. If an average of 1.0 of these is used to turn one of the plentiful U-238s into a Pu-239, then the total amount of fissile fuel in the reactor will remain unchanged. The reactor will burn a U-235, but replace it with a Pu-239, which, from a fission point of view, is just about as good. While it burns fuel, it breeds fuel!

An animal breeder can stay in business forever without buying any additional stock provides he keeps his herd fed and healthy and makes sure that at least one baby animal is born and raised to maturity for each adult he sends off to market. A breeder reactor can work the same way, except that the food for the fissile herd is fertile material that eventually becomes part of the herd. Furthermore, just as the animal breeder can grow his herd by raising more stock than he sends to market, so can the breeder reactor. If its breeding ratio is greater than 1.0, it can actually *increase* the total amount of available fissile material while it burns.

The breeder's energy yield is not infinite. To make it go, you need to keep feeding it with fertile material, which can be U-238 or Th-232. But these are respectively 130 and 500 times more plentiful than U-235. So while the breeder reactor might not provide a completely free lunch, it offers a really cheap one. To the atomic scientists of the 1940s grappling with the problem of a fictitious uranium shortage, it seemed like the necessary answer. As a result, it was assumed by many that practical commercial nuclear power would have to be based on breeder reactors, and significant research was directed accordingly.

To make a critical reactor that is not a breeder, only one neutron resulting from each fission needs to be put to good use. The rest can be wasted, and the show will still go on. But to make a breeder reactor work properly, two neutrons from each fission – one to sustain the fission and another to breed more fuel – must be properly utilized. From a practical point of view that means that the breeder reactor conditions must be such that the average number of neutrons born per fission must be substantially above 2.0. This is true for reactors fueled by Pu-239 if the neutron spectrum is fast – which can hit a batting average of as high as 3.0 if the reactor is running on unmoderated high energy MeV-class neutrons newly born from fission. That is so good that, in principle, a fast breeder reactor using

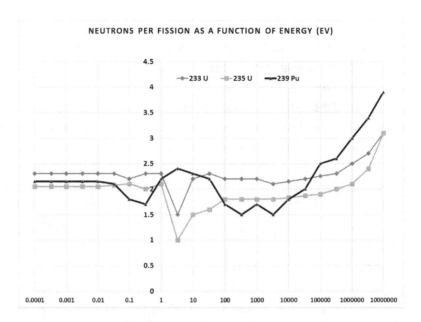

NEUTRONS PER FISSION AS A FUNCTION OF ENERGY (EV)

**Fig. 10.1** Neutrons produced per fission. In the low energy thermal ($10^{-2}$ ev) range, U-233 produces the most neutrons. In the fast ($10^6$ ev) range, Pu-239 produces the most neutrons. (Credit: Adapted from data provided Adapted and simplified from data provided by Brookhaven National Lab[2])

the U-238/Pu-239 cycle could actually produce *twice* as much fissile fuel as it burns! The fission cross section of U-235 and Pu-239 for such high energy neutrons is much smaller than it is for thermal neutrons, so for them to sustain a chain reaction, the fuel must be enriched to at least 20 percent fissile material. This is more than ten times the enrichment required for these materials to sustain criticality using highly moderated thermal neutrons. Unfortunately, however, the typical neutron yield of mixed U-235/Pu-239 under thermal conditions is only about 2.1 per event, which doesn't leave enough margin against losses to sustain a breeder reactor. In fact, thermal reactors running on low enriched uranium, such as typical PWRs, only manage to breed about 0.6 Pu-239s for every U-235 they burn. This "conversion" is still useful, it's like getting three free Pu-239s for every five U-235s you buy. But if you want to make an actual breeder reactor using the U-238 to Pu-239 cycle, it needs to be a highly enriched fast reactor.

Thorium-232 offers an alternative fertile substance to produce fissile fuel. Th-232 is radioactive, but its half-life is 4.5 billion years, making it about 1/4 as radiative as natural uranium, and four times as common. When Th-232 absorbs a neutron, it becomes protactinium 233 which quickly decays to fissile U-233. The situation with this Th-232 to U-233 cycle is the exact opposite of the U-239/Pu-239 cycle. The Th-232/U-233 combination is inferior to U-238/Pu-239 under fast reactor conditions, both because it yields fewer neutrons per fast fission than Pu-239 and the fact shown in Table 5.1 that while U-238 itself actually has an appreciable fission cross section under fast reactor conditions, Th-232 does not. However at thermal energies, U-233 can yield about 2.3 neutrons per fission, which is enough to enable a breeder reactor.

The bottom line: if you want to build a breeder reactor, there are two ways to do it. You can build a fast highly-enriched reactor using the U-238/Pu 239 cycle, or you can build a thermal low-enriched reactor using the Th-232/U-233 cycle.

While the "uranium shortage" that provided the motivation for early interest in breeder reactors proved to be fictional, breeders offer other advantages that remain valid. In particular, by reducing the amount of nuclear fuel that needs to be used to produce a given amount of energy a hundredfold, breeders can also reduce the total volume of nuclear waste a hundredfold. Furthermore non-breeding fission reactors produce two types of nuclear waste: actinides and fission fragments. Actinides are high atomic weight isotopes like uranium, neptunium, plutonium, americium, and curium that result from the absorption of neutrons by other actinides. Fission fragments are middle weight isotopes like strontium and cesium. The actinides are either fissile or fertile. In a breeder reactor they are burned as fuel, leaving only the fission fragments in the radioactive waste. This is important not just for fuel economy, but for waste disposal. The radioactive actinides have half-lives that are all over the map, including most notably Pu-239 at 24,000 years. In contrast, all the fission fragments either have half-lives of less than 91 years or greater than 200,000 years, with nothing in between. The short half-life wastes decay away quickly, while the very long half-life wastes are not very radioactive. By burning all the mid-duration actinide wastes, breeders make the problem of long duration radiative waste disposal much more tractable.

As can be seen from the data in Fig. 10.1, fast reactors make the most prolific breeders. If you want a fast reactor, you need to eliminate low atomic weight elements, like hydrogen, carbon or oxygen from its construction. Such reactors therefore employ liquid metal coolant (usually sodium, or a sodium-potassium alloy abbreviated NaK) and pure metal uranium or plutonium

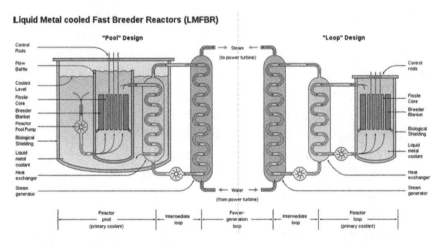

**Fig. 10.2** Pool and loop designs of liquid metal cooled fast breeder reactors. The loop design is more compact, but the pool design provides passive safety in case the cooling pumps should fail.[3] (Credit: Wikipedia Commons)

fuel. This allows them to be very compact, since liquid metal coolant has superb heat transfer properties. Furthermore, unlike high temperature water, liquid metal doesn't generate high pressure, so the massive pressure vessels employed by PWRs are not required. Furthermore, liquid metal cooled reactors can operate at much higher temperatures than PWRs, and therefore can be made to generate power more efficiently.

Liquid metal fast breeder reactors can be designed as either pump using either a "pool" design or a "loop design." The difference is shown in Fig. 10.2.

The potential advantages of the liquid metal cooled fast breeder reactors were so compelling that Argonne National Lab scientist Walter Zinn was able to convince the Atomic Energy Commission to allow him to build one to investigate the concept. (Zinn, a very talented scientist, had tried to work with Rickover, but found it impossible.[4] With Rickover there could be only one lion on a hill.) Accordingly, a reactor, dubbed Experimental Breeder Reactor 1 (EBR-1) was built at the National Reactor Testing Station (since 2005 Idaho National Lab) to validate the concept. EBR-1 was a loop design employing uranium metal fuel and NaK coolant. On December 20, 1951, the reactor went critical, generating 200 kWe of electrical power and 1.4 MW of thermal output. The following day, EBR-1's power was used to light up the building housing it, making it the first nuclear electrical power station in the world.

EBR-1 produced more plutonium fuel than it burned, proving the principle of the breeder reactor. It experienced a partial meltdown in 1955, but was repaired, and remained in operation until 1964. It was then replaced by EBR-2, a pool-type 62 MW thermal unit producing 20 MWe of electricity, sufficient to provide most of the electricity and heat to power the entire lab. In 1986, an experiment was run to test the passive safety feature of EBR-2's pool type design. With the reactor running at full power, all EBR-2's cooling pumps were simply switched

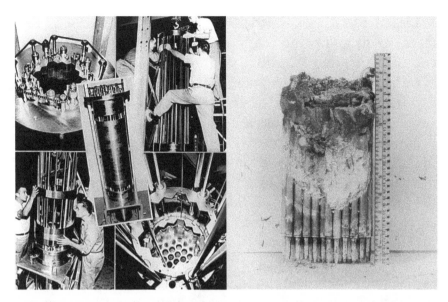

**Fig. 10.3** (Left) The core of EBR-1 being assembled in Idaho in 1951. (Right) The core after a partial meltdown even̄ in 1955. (Credits, US AEC.)

off. This caused the reactor to heat up, causing it to go subcritical, dropping the power level to the 7 percent of rated power delivered by the radioactive waste products that had accumulated in the fuel after twenty years of operation. But the natural circulation of the liquid sodium coolant pool proved sufficient to remove this decay heat. This demonstrated that the reactor was completely passively safe, able to cool itself without any active pumps or operator intervention.[5] EBR-2 continued to operate successfully until 1994, paving the way for an even larger passively safe liquid metal cooled fast breeder reactor, known as the Integral Fast Reactor. Unfortunately that program was cancelled by the Clinton administration's fanatically antinuclear Secretary of Energy Hazel O'Leary.

However, while the US Department of Energy liquid metal fast breeder program was wrecked by political vandalism, work on the concept has continued to advance internationally.

Liquid metal fast breeders have been built under government sponsorship internationally, including three in France, two in the United Kingdom, two in Russia, two in Japan, two in India, and an unknown number in China. At this writing the two Russian commercial scale fast breeders are operating and generating power in Russia, and research on more advanced designs is being avidly pursued in that country, as well as China, India, and South Korea. Even more encouragingly, as we shall discuss in the following chapter, the concept has been adopted for private development by several American and Canadian entrepreneurial organizations.[6]

But, as previously noted, there is an alternative route to the breeder reactor: the thermal thorium breeder. Such a concept can be implemented in a number of different ways. The most straightforward is just to build a conventional PWR and fuel it with a mixture of U-233 and Th-232, and let it run. This, in fact, was done by Rickover, who demonstrated precisely such an approach by converting his original 60 MWe Shippingport reactor to U-233/Th-232 in the late 1970s. Examination of the fuel after a several year run showed that more U-233 was created in a Th-232 blanket surrounding the central primary "seed" reaction region than was burned in the reactor, demonstrating a breeding ratio of about 1.01.[7] To actually make productive use of the U-233 produced in the blanket would have required fuel processing, a US capability that the Carter administration had concurrently dismantled. Nevertheless, as far as the reactor technology part of the system is concerned, employing PWRs as Th-232/U-233 thermal breeders were demonstrated.

There is, however, a more radical approach to the thermal Th-232-U-233 breeder that was investigated and demonstrated by Oak Ridge National Lab's visionary leader Alvin Weinberg in the 1950s and 1960s. This is what is known as the homogenous reactor.[8]

If you followed my instructions in Chapter 7 and built your PWR according to orthodox techniques, you were probably annoyed by the time and expense involved in fabricating thousands of fuel pellets, sticking them in hundreds of zirconium tubes, and then assembling groups of such tubes into assemblies with water cooling channels running between them and all the rest of that jazz. Weinberg wanted nuclear power to be cheap and figured that complicated stuff like that was likely to get in the way. So instead he favored developing a concept originally proposed by Manhattan Project scientists Eugene Wigner and Harold Urey in 1944 to do away with all internal structure and simply dissolve the U-235 fuel in its water moderator, and just let it boil itself. In addition to eliminating the cost of creating complex internal structures, this concept also offered the important advantage of allowing the fission product Xe-135 gas to bubble right out of solution, thereby preventing it from poisoning the reaction as had happened at Hanford during the war. Furthermore, during long-term operation other nuclear wastes could be filtered out of the solution, and replacement fuel added in as required, keeping the reactivity of the solution constant over time, without the need to shut down and replace old fuel rods with new ones. This also made the system a candidate for an optimal breeder, since there would be no need to chemically reprocess such old fuel rods to obtain their plutonium. Instead the plutonium produced in the reactor could just stay in the reactor and be used as fuel, or, if there was a surplus (i.e. breeding ratio greater than 1.0) the extra amount could be filtered out on an ongoing basis.

Weinberg's team determined that the best way to dissolve uranium in water was by using in the form of a uranyl sulfate ($^{235}UO_2SO_4$) salt. Accordingly they built Homogenous Reactor Experiment-1.

HRE-1 considered on a stainless-steel sphere 18 inches in diameter filled with a 0.17 molal solution of $^{235}UO_2SO_4$ in light water, surrounded by a 10-inch-thick blanket of heavy water. Control plates were suspended in the heavy water, enabling the blanket to act as either a neutron reflector or neutron absorber. The reactor was pressurized to 67 atmospheres, allowing it to operate with an outlet temperature of 250 C.

(This sounds pretty simple. If you know a good supplier of $^{235}$U, then instead of building a PWR according to the instructions I gave you in Chapter 7, you might make your first nuke one of these.)

On February 24, 1953, HRE-1 was turned on and brought to its full power rating of 1 MW thermal. Employing a small steam turbine, it generated 150 kWe which was fed into the TVA grid.

Weinberg describes the experience in his memoir, *The First Nuclear Era*:[9]

"This was the second nuclear reactor to generate electricity. Today, forty years later, I can still remember our feeling of elation. We youngsters at Oak Ridge (the average age of the crew in the control room was barely 30), who had been thrown out of reactor development 5 years earlier, had succeeded in one of the most difficult nuclear engineering tours de force: the stable operation of a high-power density aqueous homogenous reactor. Ed Bohhmann...silver plated several medallions to commemorate the event. I sent a medallion to Lise Meitner...

"HRE-1 operated ...for about 1,000 hours before it was shut down...What did we learn in those 1,000 hours? Most importantly we found that bubble-induced nuclear instability was eliminated by the high pressure and, a little later by the use of copper to suppress the evolution of hydrogen and oxygen. We also found that the control plates were unnecessary! We could adjust the temperature of the reactor by changing the concentration of the uranium; and because of its strong negative coefficient of reactivity, the power output matched the power demanded by the steam turbine. To shut off the reactor, we merely had to close the valve to the steam turbine.... Finally, the $^{135}$Xe came off easily."

HRE-1 was a brilliant success. But when the design was scaled up and the 5 MW HRE-2 turned on in 1958, instabilities arose due to dead regions in the flow pattern in the core tank. Fuel caught in the dead regions generated heat that could not be dissipated, causing those regions to heat up until the uranium precipitated out of solution. This created hot spots that melted the core tank in two places. The holes, however, were patched and the core flow problem corrected, stable operation at 5 MW was achieved, with 5300 hours total hours at full power including 105 days non-stop operation by 1961. But on April 18, 1961, there was another melt through the tank and the unit was shut down.

The principal problem with the aqueous homogenous reactor was that the solubility of the uranium salt in water depended upon temperature, and under unusual conditions it could precipitate out, creating local hot spots. Another problem was the high pressure the system needed to operate on. Weinberg decided to address both by replacing the water solution with one based on molten salt.

Molten salts, like LiF and BeF$_2$, sound like a daunting choice for use as a reactor fluid medium. Lithium and fluorine are both very reactive, so one might think that LiF would be quite corrosive. In fact, the opposite is true. Precisely because Li and F are so reactive, they bind tightly to each other, and neither has any inclination to waste time binding with less reactive elements like those present in steel alloys. This makes them much less corrosive than highly acidic aqueous sulfate solutions. Furthermore, unlike high temperature water, molten salts have no appreciable vapor pressure, allowing reactors employing them to operate at temperatures as high as 600 C without the need for heavy pressure vessels.

The Molten Salt Reactor Experiment was originally designed to operate at a power of 10 MW employing a molten fuel consisting of 1%$^{235}$ $UF_4$, 1% $ThF_4$, 5% $ZrF_4$, 70% $^7LiF$, and 23% $BeF_2$, which transferred its energy to a coolant salt consisting of 66%LiF, 34% $BeF_2$. The fuel flowed through a cylindrical array of graphite columns, which provided the moderation necessary to achieve criticality in such a low enriched fuel system. It began operation in early 1966 and ran smoothly until December of 1969. Towards the end of its run, the U-235 was replaced with U-233, demonstrating its ability to operate as a Th-232/U-233 breeder.

Following this brilliant success, Weinberg asked the AEC for funds to build a full-scale molten salt lithium fluoride thorium breeder reactor (LIFTR).

Unfortunately the AEC said no. With the Vietnam War pinching budgets, the AEC decided it could only support one breeder reactor program. Feeling uncomfortable with the molten salt concept, the AEC's director of reactor development Milton Shaw chose the liquid metal fast breeder as the only candidate for further development.

So the US government abandoned its LIFTR program, and as noted above, a quarter century later the liquid metal fast breeder would meet the same fate.

But if the breeder reactors had lost the battle, they had not lost the war.

America's government might be out of the game, but America's entrepreneurs were not.

# CHAPTER 11

# ENTREPRENEURIAL NUKES

**IN THE FACE** of the necessity for nuclear power made apparent by the carbon emissions problem, a number of innovative companies have been formed to try to develop a new generation of nuclear power reactors that could overcome the industry's difficulties, with many receiving serious investment money from heavy hitters like Bill Gates. The features of the new reactor designs include standardization, so that each new one shouldn't have to go through the entire licensing process; modularity, to allow them to be used to meet market needs at all scales; factory production, to make them cheaper; and passive safety, to prevent them from melting down even if all cooling systems should fail. Some use thorium fuel, with a few also employing their fuel not in solid pellets cooled by water, but in a mixture of liquid molten salts, which can pumped around to allow the reactor fuel to be continually reprocessed. This last feature would both expand the useful energy content of the fuel and cut radioactive waste by about a factor of a hundred.[1]

The new generation of nuclear reactors includes four major types, distinguished by how they are cooled. These are water cooled reactors, gas-cooled reactors, liquid metal cooled reactors, and molten salt cooled reactors. Let's have a look at each.

## 1. WATER COOLED REACTORS

The vast majority of commercial nuclear reactors around the world today are water cooled and moderated. The is a result of a decision by Admiral Hyman Rickover in the 1950s to make Pressurized Water Reactors (PWRs) the basis for the US nuclear navy. Highly successful in submarines, water reactors were the proven candidate, and thus the obvious choice, for further development as civilian power reactors. While frequently criticized over the years, Rickover's decision had plenty of logic behind it. Water can serve both as an excellent moderator and as a coolant, and, as Rickover said "has no holes in it" except those created by excessive boiling, which thereby serve to instantly stop any power excursion by weakening the moderation required to sustain the nuclear chain reaction. On the downside, in order for water to be used as a coolant at the temperatures over 300 Centigrade required to run a nuclear power plant, it has to be kept under very high pressure. This means that a water-cooled nuclear power reactor needs a very strong pressure vessel to contain the reactor, and a system of high pressure, high temperature steam piping containing moderately radioactive water (because of infusion of trace amounts of reactor fission products) to go along with it. The complexity of this plumbing system offers opportunity for error, as occurred at Three Mile Island. But it must be said that despite all the what-if accident scenarios that anti-nuclear activists have proposed for water-cooled reactors, the actual safety record of the nuclear industry has been excellent- far better in fact than all other power sources.

PWRs come in a variety of types, with the most important division among them being between light water reactors (LWRs) which use ordinary water as their moderator, and heavy water reactors (HWRs) which use heavy water, in which normal hydrogen is replaced by the hydrogen isotope deuterium,

D, or $^2$H, hydrogen whose nuclei contains both a proton and a neutron, instead of just a proton. (The Boiling Water Reactor, BWR, is an LWR that operates at a lower pressure than a standard PWR, and so generates steam directly in the core, without the need for a steam generator or secondary loop. Only a small minority of LWRs are BWRs.) Heavy water is a superior moderator to light water, because, unlike ordinary hydrogen, deuterium nuclei never absorb neutrons, but just reflect them. For this reason, HWRs, like the Canadian CANDU reactors, can actually use natural uranium as fuel, or slightly enriched uranium with a lot of thorium mixed in. LWRs, in contrast, must have their fuel enriched from the 0.7% U-235 present in natural uranium to about 3% U-235. Such "reactor grade" fuel, however, is cheap enough; it only represents about 5% of the cost of power generated by LWRs, a fact that has limited the appeal of HWRs, breeder reactors, and thorium fuel (which can be, and has been, used to a lesser degree as a component of the fuel in LWRs).

So rather than seek innovations in the fuel cycle, designers of the next generation of LWRs are aiming at what they see as the heart of the matter: reducing capital costs. They hope to do this by making the reactors smaller, 50 to 300 MWe each, instead of the 1000 MWe units typical of existing commercial LWRs. This will allow such small modular light water reactors (SMRs) to be massed produced according to a standard design, which will be licensed well in advance of all utility projects that use them.[3] Furthermore, as studies have shown that it takes an average of eight times as long to do anything, say a weld, at a construction site as it does in a factory, the units will be basically factory made, and merely assembled in site. Thus, licensed in advance, and mass produced in factories using a standardized design with interchangeable parts, such reactors should enable nuclear power stations to be built much more quickly and cheaply, with much less opportunity for malicious litigation interference than is currently the case.

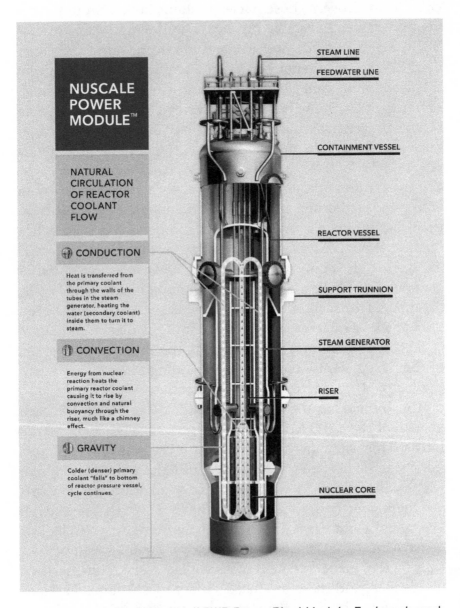

**Fig. 11.1** NuScale Modular Small PWR Power Plant Module. Each such module produces 77 MWe. A large water pool surrounds groups of 77 MW reactor units.[2] (Credit: Wikipedia Commons)

The NuScale company is generally considered a leader in the development of SMRs. Their design consists of a group of up to twelve 77 MWe units, all drawing their water from a common pool. It also includes a variety of passive safety systems that need electric power to *prevent activation*. If electric power to the system is cut off for any reason, they kick in automatically, shut the reactor down, and keep it cooled.[4]

## 2. HIGH TEMPERATURE GAS-COOLED REACTORS

Graphite can also be used to moderate nuclear reactors, and in fact the first nuclear reactor ever built, that constructed by Enrico Fermi and his Manhattan Project team beneath the stands of the football stadium at the University of Chicago in 1942, was graphite moderated. Graphite does not have the unstoppable negative temperature reactivity coefficient that is built in, as a law of physics, in water moderated reactors. But with proper design, they can be temperature-reactivity negative. (Unfortunately, the graphite-moderated Chernobyl reactor was not properly designed.) A small number of commercial graphite-moderated gas-cooled reactors have been built and operated in the West, including one at Fort St. Vrain, Colorado.[5]

The big safety advantage of graphite moderated reactors is that they cannot melt down, because, unlike the zirconium cladding of the uranium fuel pellets in the water-cooled Three Mile Island reactor, which could and did melt after cooling was shut off, graphite simply doesn't melt. In next generation graphite reactors the uranium fuel is contained as tiny bits within small poppy seed sized "TRISO" particles made of unmeltable carbon and silicon carbide. These in turn will be contained either in graphite blocks or graphite billiard balls, which will be cooled by helium gas. If the cooling is ever cut off, the chain reaction will be shut down by the negative temperature reactivity coefficient, and while there will be decay heat (initially amounting to about 7% the rated power of the reactor) from radioactive waste products scattered within the fuel bits inside the TRISO particles, the graphite will be tough enough to simply withstand it.[6] When water cooling was cut off to the Three Miles Island reactor, the zirconium cladding of its uranium fuel pellets melted down, destroying the unit. If it had been a graphite reactor, nothing would have happened at all.

In addition to safety, however, gas cooled graphite reactors offer another advantage over water cooled reactors: they can be run much hotter. The hottest a water-cooled reactor can ever run is about 350 C, because if you try to run much hotter than that the pressure needed to keep the water liquid becomes infinite. In contrast, high temperature gas-cooled reactors (HTGRs) graphite moderated reactors can be run as hot as 700 C, and advanced designs are in the works for very high temperature gas cooled reactors (VHTGRs) that may be able to run as hit as

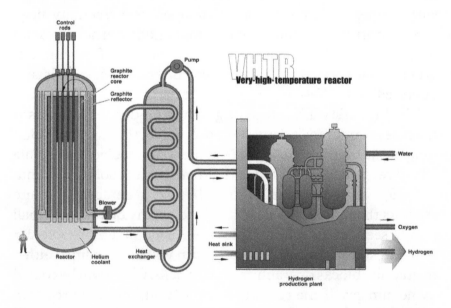

**Fig. 11.2** Very High Temperature Reactor Hydrogen Generating Plant (Credit: US Department of Energy)

1000 C.[7] Achieving such high temperatures is of great benefit because the high the reactor temperature the more efficient the steam cycle will be used to convert the reactor's thermal power to electricity. Existing LWRs run at about 33% efficiency. HTG-CRs should be able to run at over 50% efficiency, and VTGCRs at 65 percent. Very high temperature reactors can also be used to directly produce hydrogen, which could then be utilized as a fuel for zero emission cars.

Next generation HTGRs are currently being developed by a number of companies, including Framatome, X-energy, U-Battery, General Atomics, and STARCORE. One is also under construction in China.

## 3. LIQUID METAL COOLED REACTORS

As noted, one of the downsides of using water as a coolant is that it has to be kept under very high pressure, and even so, the maximum temperature of any water-cooled reactor is limited. For this reason some early reactor designers hit on the idea of using liquid metal as a coolant. So the second US Navy nuclear submarine, the *Seawolf*, used a reactor cooled by liquid sodium. (The first, in the *Nautilus*, was an LWR.) Sodium melts at 98 C and is a terrific coolant because of its high thermal conductivity, but as it explodes on contact with water, it poses a number of scary drawbacks for submarine propulsion. After the *Seawolf* experienced some problems with sodium leaks, Rickover ultimately decided to go forward with LWRs instead. The Soviet navy, on the other hand, felt that the compact nature of liquid metal cooled reactors outweighed such issues, and took them further, outfitting many of its submarines with reactors cooled by liquid lead.[8]

Liquid metals make poor moderators, and so liquid metal reactors are what is known as "fast" reactors, because they sustain their chain reaction with fast neutrons, newborn from

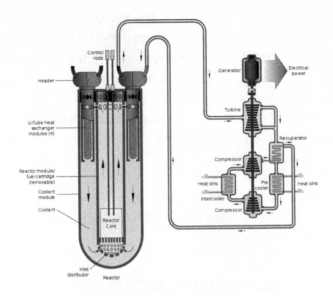

**Fig. 11.3** Lead Cooled Fast Reactor (Credit: US Department of Energy)

fission, rather than slow moving neutrons that have been moderated down to "thermal" (i.e. low) velocities. Fast neutrons are less reactive than thermal neutrons, and so fast reactors must use fuel enriched to 20% fissile material or better. An advantage of fast reactors is that fast neutrons are not subject to parasitic absorption by nuclear waste products as easily as thermal neutrons are, so they can use a lot more of their uranium before the accumulation of waste products in the fuel ruins it for them. In fact, they can actually burn up most of the worst nuclear waste products. Metal cooled reactors can also use metallic uranium fuel, which expands when it is heated, creating strong negative temperature reactivity feedback. They can also operate at higher temperatures than LWRS, about 550 C, offering a significant improvement in efficiency, albeit not as much as that available from HTGR designs. Because they keep the uranium fuel in metal form, metal cooled reactors are attractive for use

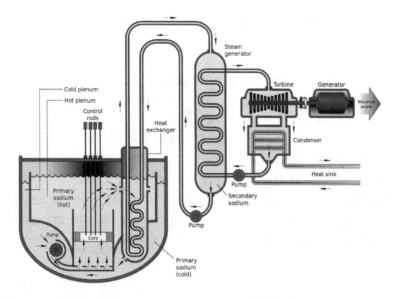

**Fig 11.4** Sodium Cooled Fast Reactor (Credit: US Department of Energy)

as breeder reactors, recycling plutonium-239 produced by the absorption of neutrons in uranium-238 back into the reactor (or into other reactors) as fuel. Such a liquid metal cooled fast breeder reactor, called the Superphoenix, actually operated in France for a number of years, before it was shut down due to a variety of mechanical problems.

Small modular fast metal cooled reactors are currently under development by Westinghouse, General Electric, Oklo, and Terrapower.

## 4. MOLTEN SALT COOLED REACTORS

Liquid metals are not the only choice for a high temperature, low pressure, liquid reactor coolant. An alternative is to use liquid salts. In these molten salt reactors (MSRs), the fuel can be dissolved right into the coolant, delivering its heat directly. A catastrophic meltdown of such a reactor is impossible, because, well, it's already happened. In an MSR, a meltdown isn't a bug, it's a feature!

Molten salts have some moderating capability, and MSRs may add to this by having some graphite in and around the reactor vessel, giving them a medium-energy ("epithermal") neutron spectrum. As a result their fuel needs to be enriched to a level beyond that of LWRs, but less than metal cooled reactors. The hot liquid salt/fuel mixture is pumped around and delivers its heat to water or another turbine working fluid at temperatures as high as 650 C, with 1000 C possible provided materials corrosion problems at higher temperatures can be overcome.

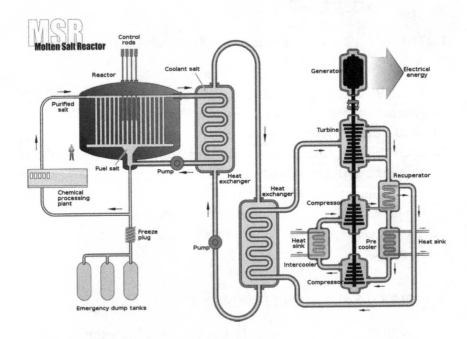

**Fig. 11.5** Molten Salt Reactor. The molten fissile salt in the primary loop can be continually reprocessed, eliminating fission product poisons and allowing maximum fuel burnup and breading ratios, with no need to shut down. (Credit: US Department of Energy)

Because the salt/fuel mixture is constantly being pumped out of the reactor, it can be continually reprocessed, eliminating nuclear wastes like Xe-135 that poison the fuel in other reactor systems. As a result an MSR can achieve much higher fuel burnup than other reactor types – about a hundred times as great, with the amount of nuclear waste created for a given amount of power generated reduced a hundredfold as well. This feature also makes the MSR a very attractive design for a breeder reactor, using thorium to breed U-233 or uranium-238 to breed plutonium to fuel other reactors, without ever needing to shut down to remove spent fuel for reprocessing.

It was for this latter reason that the MSR was proposed and championed by Alvin Weinberg, the visionary director of what is now known as Oak Ridge National Lab (ORNL) in 1960s. Under Weinberg's leadership a MSR was built and tested, running quite successfully from 1965 to 1969. However the program ran afoul of other Atomic Energy Commission (now DOE) leaders, who decided to limit the US breeder effort to liquid metal fast reactors.[10]

MSRs today are being actively pursued by a number of companies, including the US based ThorCon and Southern company/Terrapower, Britain's Moltex company, and Canada's Terrestrial Energy.

With such designs, there can be little rational basis for opposition to nuclear power. In fact, a number of center-left organizations, notably the Breakthrough Institute and the important Third Way group within the Democratic Party, being very concerned about global warming and environmental damage, have embraced the new generation of advanced nuclear power reactors as offering the cleanest, safest, most abundant, potentially cheapest, and far and away the most environmentally benign energy option available. If they can prevail in their effort to convince Progressives to embrace practical, rather than ideological, forms of environmentalism, the possibilities for a nuclear renaissance in the West will improve radically.

That, *if,* in my view, is the fundamental question. There really is no problem with existing LWRs. Since the 1950s, over a thousand of these units have operated on land and sea, without a single casualty among either operators or the public from any radiological release. That track record makes them by far the safest energy source ever devised. Practical environmentalists support nuclear energy because it is a safe way to produce unlimited amounts of energy with minimal pollution or harm to nature. Ideological environmentalists oppose it for precisely that reason. In fact – despite coal combustion fumes blackening the air and oil spills despoiling coastlines – they fight it with far greater vehemence than they do fossil fuels. This is so because air and water pollution are not threats to them. Far from it, they are fundraisers. They hate nuclear power because it solves problems they need to have.

Unless practical environmentalists can discredit and dethrone these types, no amount of reactor redesign will make a difference.

But whether they do or they don't, nuclear power is coming, bigtime. It is the energy source that will power the human future, although unfortunately America might not be part of it. There are about 450 nuclear reactors in the world today. By 2050, China intends to build 450 more, domestically, and is actively vying with Russia for hundreds of more nuclear projects that will be built in the developing sector. It may take 16 years to build an LWR in the USA, but it still takes only 4 years to build one in South Korea. So maybe the South Koreans, or the Indians, will offer some free world competition to the global nuclear renaissance. The US could too – and win with some of these more advanced designs. But it will take a societal change of heart to make it so.

## FOCUS SECTION: NEW NUCLEAR PLANTS UNDER CONSTRUCTION

While Americans and Europeans may think that nuclear power is a declining industry, this is far from true. About 60 reactors are currently under construction worldwide. Here is a list of 55 of them. The list includes three Indian CANDU-type pressurized heavy water reactors (PHWR), one Indian, 1 Russian, and two Chinese liquid metal cooled fast breeder reactors (FBR, BREST-300, CFR600). The other 48 units are all PWRs. About 80 percent of the units are being built in Asia. The list includes 2 American AP1000 reactors, 4 South Korean AP1400s, three French EPRs, 1 German Pre-Konvoi, 21 Russian VVERs, 1 Argentine Carem25, and three Indian PHWRs. The other 20 reactors are all Chinese.[11]

| Start † | | Reactor | Model | Gross MWe |
|---|---|---|---|---|
| 2023 | China, CGN | Fangchenggang 3 | Hualong One | 1180 |
| 2023 | Belarus, BNPP | Ostrovets 2 | VVER-1200 | 1194 |
| 2023 | Bangladesh | Rooppur 1 | VVER-1200 | 1200 |
| 2023 | China, CNNC | Xiapu 1 | CFR600 | 600 |
| 2023 | India, NPCIL | Kakrapar 4 | PHWR-700 | 700 |
| 2023 | India, NPCIL | Rajasthan 7 | PHWR-700 | 700 |
| 2023 | India, NPCIL | Rajasthan 8 | PHWR-700 | 700 |
| 2023 | Korea, KHNP | Shin Hanul 2 | APR1400 | 1400 |
| 2023 | Korea, KHNP | Shin Kori 5 | APR1400 | 1400 |
| 2023 | Slovakia, SE | Mochovce 3 | VVER-440 | 471 |

| Start † | | Reactor | Model | Gross MWe |
|---------|---|---------|-------|-----------|
| 2023 | Turkey | Akkuyu 1 | VVER-1200 | 1200 |
| 2023 | UAE, ENEC | Barakah 4 | APR1400 | 1400 |
| 2023 | USA, Southern | Vogtle 3 | AP1000 | 1250 |
| 2023 | USA, Southern | Vogtle 4 | AP1000 | 1250 |
| 2024 | Slovakia, SE | Mochovce 4 | VVER-440 | 471 |
| 2024 | India, NPCIL | Kalpakkam PFBR | FBR | 500 |
| 2024 | France, EDF | Flamanville 3 | EPR | 1650 |
| 2024 | Bangladesh | Rooppur 2 | VVER-1200 | 1200 |
| 2024 | China, CGN | Fangchenggang 4 | Hualong One | 1180 |
| 2024 | China, SPIC & Huaneng | Shidaowan 1 | CAP1400 | 1500 |
| 2024 | China, Guodian & CNNC | Zhangzhou 1 | Hualong One | 1212 |
| 2024 | Iran | Bushehr 2 | VVER-1000 | 1057 |
| 2024 | Korea, KHNP | Shin Kori 6 | APR1400 | 1400 |
| 2024 | Turkey | Akkuyu 2 | VVER-1200 | 1200 |
| 2025 | India, NPCIL | Kudankulam 3 | VVER-1000 | 1000 |
| 2025 | India, NPCIL | Kudankulam 4 | VVER-1000 | 1000 |
| 2025 | Russia, Rosenergoatom | Kursk II-1 | VVER-TOI | 1255 |
| 2025 | Russia, Rosenergoatom | Kursk II-2 | VVER-TOI | 1255 |

| Start † | | Reactor | Model | Gross MWe |
|---------|---|---------|-------|-----------|
| 2025 | China, SPIC & Huaneng | Shidaowan 2 | CAP1400 | 1500 |
| 2025 | China, CGN | Taipingling 1 | Hualong One | 1200 |
| 2025 | China, Guodian & CNNC | Zhangzhou 2 | Hualong One | 1212 |
| 2025 | Turkey | Akkuyu 3 | VVER-1200 | 1200 |
| 2026 | China, CGN | Cangnan/San'ao 1 | Hualong One | 1150 |
| 2026 | China, Huaneng & CNNC | Changjiang 3 | Hualong One | 1200 |
| 2026 | China, CNNC | Changjiang SMR 1 | ACP100 | 125 |
| 2026 | China, CGN | Taipingling 2 | Hualong One | 1202 |
| 2026 | China, CNNC | Tianwan 7 | VVER-1200 | 1200 |
| 2026 | China, CNNC | Xiapu 2 | CFR600 | 600 |
| 2026 | Russia, Rosatom | BREST-OD-300 | BREST-300 | 300 |
| 2026 | Turkey | Akkuyu 4 | VVER-1200 | 1200 |
| 2027 | India, NPCIL | Kudankulam 5 | VVER-1000 | 1000 |
| 2027 | Argentina, CNEA | Carem | Carem25 | 29 |
| 2027 | China, CGN | Cangnan/San'ao 2 | Hualong One | 1150 |
| 2027 | China, CNNC | Sanmen 3 | CAP1000 | 1250 |
| 2027 | China, CNNC | Tianwan 8 | VVER-1200 | 1200 |

| Start † | | Reactor | Model | Gross MWe |
|---|---|---|---|---|
| 2027 | China, CNNC & Datang | Xudabao 3 | VVER-1200 | 1200 |
| 2027 | China, Huaneng & CNNC | Changjiang 4 | Hualong One | 1200 |
| 2027 | India, NPCIL | Kudankulam 6 | VVER-1000 | 1000 |
| 2027 | UK, EDF | Hinkley Point C1 | EPR | 1720 |
| 2028 | Brazil, Eletrobrás | Angra 3 | Pre-Konvoi | 1405 |
| 2028 | China, CGN | Lufeng 5 | Hualong One | 1200 |
| 2028 | China, CNNC & Datang | Xudabao 4 | VVER-1200 | 1200 |
| 2028 | Egypt, NPPA | El Dabaa 1 | VVER-1200 | 1200 |
| 2028 | UK, EDF | Hinkley Point C2 | EPR | 1720 |
| 2030 | Egypt, NPPA | El Dabaa 2 | VVER-1200 | 1200 |

† *Latest announced/estimated year of grid connection.*
*Note: units where construction is currently suspended are omitted from the above Table. These include two in Ukraine and two in Japan.*

# CHAPTER 12

# THE POWER THAT LIGHTS THE STARS

**WHILE NUCLEAR FISSION** has radically expanded humanity's potential energy resources, the achievement of controlled nuclear fusion will make them virtually infinite. The basic fuel for fusion is deuterium, an isotope of hydrogen that is called "heavy" because, in addition to having the proton in its nucleus that all hydrogen atoms have, it also has a neutron, which doubles its weight. Deuterium is found naturally on Earth; one hydrogen atom out of every 6,000 is deuterium. This number might not seem like much, but because of the enormous energy released when a fusion reaction occurs, it's enough to endow each and every gallon of water on Earth, fresh or salt, with a fusion energy content equivalent to that obtained by burning 350 gallons of gasoline.[1] The Earth's fusion resources are millions of times greater than its fission resources, and billions of times greater than its known fossil-fuel resources. Even at ten times our current rate of consumption, there is enough fusion fuel on this planet (alone) to power our civilization until the sun dies.

Furthermore, fusion produces no greenhouse gases, and if done correctly, need not produce significant radioactive waste. When they collide, the deuterium nuclei fuse to form tritium or helium-3 ($^3$He) nuclei, plus some neutrons. The tritium and

$^3$He will then react with other deuterium nuclei to produce ordinary helium ($^4$He) and common hydrogen ($^1$H), plus a few more neutrons. If the reactor is made of conventional materials, like stainless steel, the neutrons can produce some activation, resulting in the production of about 0.1 percent the radioactive waste of a fission breeder reactor. However, if specially chosen structural materials like carbon-carbon graphite are used, there will be no activation, and the system can produce endless amounts of energy without radioactive waste or pollution of any kind.[2]

Once we have fusion, we will be able to make as much liquid chemical fuel as we desire. For example, methanol, an excellent vehicle fuel, can readily be manufactured inorganically, simply by reacting carbon dioxide with electrolysis-produced hydrogen over copper on zinc oxide catalyst.[3] Using fusion power we could manufacture unlimited supplies of methanol from $CO_2$ and water. Under such circumstances, the OPEC nations' possession of the world's oil reserves would give them as much influence over the human future as they currently derive from their monopoly of camel milk.

But fusion is not just a plentiful source of energy; it is a new *kind* of energy. Fusion offers the potential to do things that are simply impossible without it. If we can get fusion, we will be able to use the superhot plasma that fusion reactors create as a torch to flash any kind of rock, scrap, or waste into its constituent elements, which could then be separated and turned into useful materials. Such technology would eliminate any possibility of exhausting Earth's resources. And using fusion power, we will be able create space propulsion systems with exhaust velocities hundreds or thousands of times greater than the best possible chemical rocket engines. With such technology, the stars would be within our reach.[4]

So the fusion game is really worth the candle. It's a tough game, though, because while fusion occurs naturally in the stars, creating the conditions on Earth to allow it to proceed in a controlled way, in a human-engineered machine, is quite a challenge.

All atomic nuclei are positively charged, and therefore repel each other. In order to overcome this repulsion and get nuclei to fuse, they must be made to move very fast while being held in a confined area where they will have a high probability of colliding at high speed. Superheating fusion fuel to temperatures of about 100 million degrees Celsius gets the nuclei racing about at enormous speed. This is much too hot to confine the fuel with a solid chamber wall—any known or conceivable solid material would vaporize instantly if brought to such a temperature. However, at such temperatures, matter exists in a fourth state, known as plasma, in which the electrons and nuclei of atoms move independently of each other. (In school we are taught that there are three states of matter: solid, liquid, and gas. These dominate on Earth, where plasma exists only in transient forms in flames and lightning. However, most matter in the universe is plasma, which constitutes the substance of the sun and all the stars.) Because the particles of plasma are electrically charged, their motion can be affected by magnetic fields. Magnetic traps such as the toroidal or "doughnut-shaped" tokamak (as well as a variety of alternative concepts like stellarators, magnetic mirrors, and so on) have been

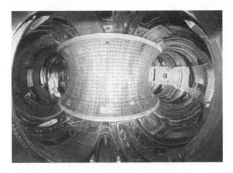

**Fig. 12.1:** Interior of the TFTR tokamak at the Princeton Plasma Physics Lab. Built in 1980, it approached breakeven, producing a world-record 10.7 megawatts of fusion power in 1994. It was shut down in 1997. No subsequent large American tokamak has been built. (Credit: Princeton Plasma Physics Laboratory)]

designed to contain fusion plasmas without ever letting them touch the chamber wall.

At least, that is how it is supposed to work in principle. In practice, all magnetic fusion confinement traps are leaky, allowing the plasma to gradually escape by diffusion. When the plasma particles escape, they quickly hit the wall and are cooled to its (by fusion standards) very low temperature, thereby causing the plasma to lose energy. However, if the plasma is producing energy through fusion reactions faster than it is losing it through leakage, it can keep itself hot and maintain itself as a standing, energy-producing fusion "fire" for as long as additional fuel is fed into the system. The denser a plasma is, and the higher its temperature, the faster it will produce fusion reactions, while the longer the individual particles remain trapped, the slower will be the rate of energy leakage. Combining these factors, it turns out that the critical factor (known as the Lawson parameter) affecting the performance of fusion systems is the product of the plasma density, the temperature, and the average particle confinement time achieved in a given machine.

As shown by the curve of binding energy presented in Chapter 4, the fusion of any elements substantially lighter than iron (atomic weight 56) can be made to produce power. However certain reactions are easier to drive then others.

The most important of the feasible fusion reactions are:

$$D + T => {}^4He + n + 17.6 \text{ MeV} \tag{12.1}$$

$$D + D => T + p + 4.0 \text{ MeV} \tag{12.2}$$

$$D + D => {}^3He + n + 3.3 \text{ MeV} \tag{12.3}$$

$$D + {}^3He => {}^4He + p + 18.3 \text{ MeV} \tag{12.4}$$

$$ {}^6Li + D => 2\,{}^4He + 22.4 \text{ MeV} \tag{12.5}$$

$$ {}^7Li + p => 2{}^4He + 17.2 \text{ MeV} \tag{12.6}$$

$$ {}^{11}B + p => 3{}^4He + 8.7 \text{ MeV} \tag{12.7}$$

The reaction cross section for the fusion reactions that are easiest to drive is shown in Fig. 12.2.

Fusion cross sections vary as a function of the plasma temperature. In a realistic fusion device, a cross section on the order of 2 x$10^{-2}$ barns (2 x$10^{-30}$ m²) will be needed to make the reaction go. This being the case, we can see that we need a plasma temperature of about 10 keV (1 keV = 10 million C) to drive the D-T reaction, 40 keV to drive the D-D reaction, 60 keV to drive the D-³He reaction, and 140 keV to drive the p-¹¹B reaction. The D-⁶Li reaction, not shown in Fig. 12.2, is about as difficult to ignite as p-¹¹B.

Tritium is very rare on Earth. But a fusion reactor could be operated as a D-T system, breeding its tritium as it goes by reacting the neutrons it emits with a lithium blanket surrounding the reactor vessel. (When a lithium nucleus absorbs a neutron, it splits into a helium and a tritium atom, and sometimes emits a neutron, which allows yet another tritium atom to

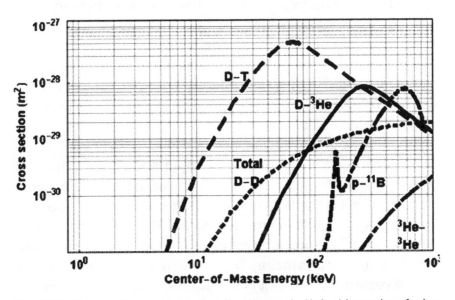

**Fig. 12.2.** The reaction cross section for the most attainable nuclear fusion reactions. ($10^{-28}$ m² = 1 barn) (Credit: US Department of Energy)

be produced.) First-generation fusion reactors may be designed along these lines.[5] However, with a little further progress in improved magnetic confinement, this will become unnecessary. Instead, once D-T ignition is reached at 10 keV, we will be able to use the plasma's own power to increase its temperature to over 40 keV at which the D-D fusion reaction and its T and $^3$He byproducts will burn by themselves.[6]

The D-T reaction gives off 20% of its energy as charged particles and 80% as neutrons. If all of its side branches are added up the D-D reactions (equations 12.2 and 12.3, which occur with equal probability) give off 60% of their energy as charged particles and 40% as neutrons. The rest of the reactions listed above, D-$^3$He, D-$^6$Li, p-$^7$Li, and p-$^{11}$B reactions are all aneutronic (although some neutron-producing D-D side reactions will occur in a D-$^3$He plasma, and – because D-D is much easier to light than D-$^6$Li – dominate in a D-$^6$Li plasma). Aneutronic reactions give off all of their energy as charged particles, and so produce zero radioactive waste. The easiest of these to drive is the D-$^3$He reaction, which can ignite around 60 keV, so any reactor that can burn D-D can also burn this fuel combination. However $^3$He does not exist naturally on Earth. It can be found on the Moon and in the atmospheres of the giant planets. It can also be produced by making tritium, which will decay into $^3$He with a half-life of 11 years.

But if we can go further, using the D-T, D-D, and D$^3$He fusion energy released to drive temperatures up to the 150 keV range, running reactors on the proton-boron will become possible.[7] This is the holy grail of energy, because protons (i.e. ordinary hydrogen nuclei) and boron-11 (which comprises 80 percent of natural boron) are extremely common, there are no neutron-producing D-D side reactions, and no isotope separation is necessary for their production. Hydrogen is the most prevalent element in the universe, while boron is about as common as copper, zinc, or lead. Once we perfect proton-boron fusion, humanity's energy resources will be beyond reckoning.

The crucial trait of a useful fusion reaction is that it produces more power than the amount of external power applied to heat the plasma (via microwave heaters, electromagnetic induction, or other means). In other words, a fusion reactor becomes viable only at the point that it produces as much power as it uses—a condition known as "breakeven." The fusion reaction in which this condition is most easily achieved is the deuterium-tritium (D-T) reaction, which for breakeven requires a Lawson parameter of $2 \times 10^{21}$ keV-s/m$^3$ (where the units are kiloelectronvolt-seconds per cubic meter). This point was nearly reached at the Japanese JT-60U tokamak in 2004. A further crucial condition, known as "ignition," occurs when the reaction becomes so powerful that it heats itself, and external heating is no longer required. For D-T, this requires a Lawson parameter of $6 \times 10^{21}$ keV-s/m$^3$. For D-D and D-$^3$He fusion it is about five times higher, and for p-$^{11}$B fusion ten times higher still. Yet, given the five orders of magnitude increase in the Lawson parameter achieved since the 1960s, reactors capable of p-$^{11}$B should ultimately be attainable.

As can be seen in Figure 12.3, the world's fusion programs have made enormous strides over the period from the 1960s through the 1990s, raising the achieved Lawson parameter by a factor of almost a hundred thousand to nearly reach breakeven. Another factor of four, which could be accomplished *if* funds are provided to build the next generation of experimental tokamaks, would take us to ignition.

Fusion can be developed, and when it is, it will eliminate the specter of energy shortages for millions of years to come. It is therefore the greatest nightmare of the Malthusians.[8] The technological challenges of fusion are significant, but provided that human ingenuity remains free, there can be little doubt that it can, and will, solve them all. There need be no limits to human aspirations, because fundamentally our wherewithal does not come from the Earth, but from ourselves. We are the ultimate resource.

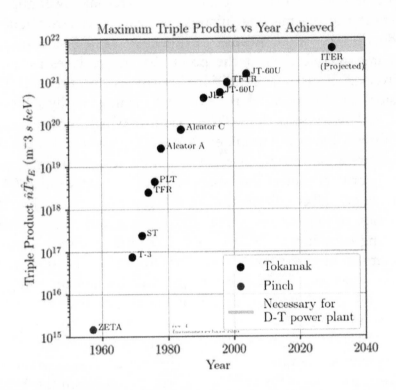

**Figure 12.3:** Progress in controlled fusion. Since 1965, the world's fusion programs have advanced the achieved Lawson parameter by a factor of 100,000. A further increase of a factor of 4 will take us to ignition. This awaits the building of new experimental machines, which has not been done since the 1990s. Note the logarithmic scale. (Credit: Fusion Energy Base.)

## HISTORY OF FUSION RESEARCH

The road to fusion was opened when the great British scientist Sir Arthur Eddington demonstrated the experimental proof of Einstein's Theory of Relativity in 1919. Shortly afterwards, in 1920, F.W. Aston published his curve of binding energy (Fig. 4.1), which showed that mass is lost when the lightest elements are fused into heavier ones. Virtually immediately, Eddington realized that according to Einstein's $E=mc^2$ formula, vast amounts of energy must be released when this occurs. This process was the answer to the mystery of what lights the stars! Furthermore, it could potentially provide a nearly infinite energy source for human civilization. In a speech to the British Association for the Advancement of Science in August 1920 he said:

"If, indeed, the subatomic energy in the stars is being freely used to maintain their great furnaces, it seems to bring a little nearer to fulfillment our dream of controlling this latent power for the well-being of the human race – or for its suicide."[9]

Further discoveries followed quickly. In 1928, Soviet (later American – he defected in 1932) physicist George Gamow worked out the quantum physics that allows atomic nuclei to fuse. This process, however, only occurs when the nuclei are brought very close together, which requires a lot of energy to overcome the electrostatic repulsion of like-charge nuclei. Shortly afterwards, Gamow visited the Cavendish Laboratory at Cambridge, and explained his theory to the young British physicist John Cockcroft. Cockcroft and his colleague Ernest Walton decided to test the theory by building an accelerator. In 1932 they put it to work, and firing hydrogen nuclei (i.e. protons) into lithium targets, observed the production of helium nuclei with energies a hundred times greater than that of the incident protons. This could not be a practical energy source, because only an infinitesimal fraction (about one in a hundred million) of the proton projectiles fired actually struck home, so a million times more energy was consumed by the accelerator than that yielded by the reactions it produced. But it was nuclear fusion.

In 1908 Ernest Rutherford had won the Nobel Prize for his discovery of the electron, a breakthrough he followed up in 1911 by laying out the basic theory that atoms consisted not of little hard balls, but dense positively charged nuclei orbited by negatively charged light electrons. By the thirties, this giant of physics was the director of the Cavendish Lab. Taking an interest in Cockcroft and Walton's fusion experiments, he expanded and continued the program, and as the thirties wore on, demonstrated several other important fusion processes, including the critical D-D reactions 12.2 and 12.3, producing both tritium and helium-3.

By 1938 enough data had been produced to allow Gamow and Hungarian-Jewish émigré physicist Edward Teller, working together at George Washington University, to conclude that the Sun's energy came from deuterium fusion.

The theory to explain how this occurred was then worked out by German-Jewish émigré physicist Hans Bethe in a series of papers published in 1938-39. First protons would fuse to form deuterium, then the deuteriums would fuse to form tritium or helium-3, which would then react to form normal helium-4. The proton-proton reaction was slow, the rest of the reactions were much faster. Therefore the proton-proton reaction determined the pace of the system, with its deuterium, tritium, and helium-3 products being reacted away into stable helium-4 quickly after they are formed. Thus the vast majority of the sun's observable composition is either ordinary hydrogen (i.e. protons) or helium-4. Bethe also identified another fusion reaction series, by which protons successively fuse with carbon to form nitrogen, then with nitrogen to form oxygen, then with oxygen to form a carbon and a helium. This "CNO cycle" dominates fusion reactions in large superhot stars, while small and medium-sized stars, like our Sun, primarily employed the proton-proton series.

Shortly afterwards, the war broke out, and all the great émigré atomic scientists (Szilard, Fermi, Gamow, Teller, Bethe, among many others) went to work alongside their American and British colleagues to produce the fission bomb. But some among them, notably Teller, kept thinking about fusion.

(The incompatibility of racial exclusivity with top-flight science cannot escape observation. The Kaiser's relatively tolerant Germany was only able to wage World War I by making use of the inventions of German-Jewish chemist Fritz Haber. Haber invented the method of artificial nitrogen fixation. This was essential to produce both ammunition and fertilizer once the British Navy had cut off Germany's access to Chilean natural nitrates. Without the Haber process, one year into the war Germany's army would have had no ammunition and its people would have had no food. Haber also invented poison gas; a secret weapon that would have won Germany the war had the German General Staff taken intelligent advantage of the surprise shock of its first employment. In contrast, the Third Reich drove Europe's best scientific talent to Los Alamos.)

In the Manhattan Project, methods of using chemical explosives to suddenly implode a mass of uranium plutonium to set off a supercritical fast chain reaction were developed. Working as part of the Manhattan Project team, Teller realized that an analogous trick could be done using a fission bomb to generate the extreme temperatures and pressures necessary to implode and ignite a mass of fusion fuel. This could produce an explosion hundreds or even thousands of times more powerful than that of the fission device used to trigger it.

Because fission bombs were necessary to trigger such "hydrogen bombs," they needed to be developed first, and since the fission bombs by themselves proved quite sufficient to end the war, the fusion bombs had to wait until afterwards to make their appearance. But they didn't need to wait long. The United States tested its first small hydrogen bomb (220 kiloton yield, ten times the ~20 kiloton Hiroshima bomb yield) using liquid deuterium fuel in 1951, and a much larger (10,000 kiloton) liquid deuterium bomb in 1952.[10]

Liquid deuterium has to be stored at 20 K, making it unworkable for use as a weapon. The Soviet program, led by Igor Kurchatov, Andrei Sakharov, Igor Tamm, Vitaly Ginzburg, Yakov Zeldovich, and Viktor Davidenko solved this problem by setting off a 400-kiloton bomb using D-$^6$Li fuel (which is a solid at room temperature) in 1953. This system works well in a combined fission-fusion bomb system because the $^6$Li combines with neutrons from the fission explosion and turns into tritium and $^4$He. This creates tritium fuel on the spot for the most readily ignitable D-T reaction. Picking up on this approach, the US tested a 15,000 kiloton D-$^6$Li bomb in 1954, while the Soviets went on to improve their thermonuclear D-$^6$Li weapon yields to 1,600 kilotons in 1955 and perform a terrifying (if impractical) 50,000 kiloton explosion demonstration in 1961.

With the secret out, other countries followed, with the UK detonating its first thermonuclear fusion bomb in 1957, China in 1967, and France in 1968.

The world's major powers spared no expense in developing hydrogen bombs, because their unlimited destructive potential made successful aggression against any country that possessed them impossible. Thus, while for the past seven decades would-be conquerors have amused themselves with a plenitude of limited conflicts in remote areas, the facts that (a) they could not possibly win, and (b) they would be killed themselves, has deterred all such gentlemen from initiating a general war.

Thus, in contrast to the previous century and half which featured an all-out war between major contestants roughly every ten to fifteen years (the French Revolutionary and

Napoleonic Wars 1789-1815, the Greek War of Independence/ Russo-Turkish War 1821-1832, the Mexican-American War 1846-1848, the Crimean War 1853-1856, the Taiping War 1850-1864, the American Civil War 1861-1865, the Austro-Prussian War 1866, the Franco-Prussian War 1870-1871, the Russo-Turkish War 1877-1878, the Sino-Japanese War 1894-1895, the Spanish-American War 1898, the Russo-Japanese War 1904-1905, World War I 1914-1918, the Russian Civil War 1917-1923, the Japanese Invasion of Manchuria 1931-1932, the Second Sino-Japanese War 1937-1945, World War II 1939-1945, and the Korean War 1950-1953) the past seventy years have, for the most part, been ones of general peace.

It is true while I write these lines in December 2022 Russia has invaded Ukraine. *But this was only possible because Ukraine gave up its nuclear arsenal.* Meanwhile, the nuclear-armed United States and its NATO allies have remained untouched, and except for a severe winter storm that has snarled things in the US, and an energy shortage caused by Europeans' absurd decision to shut down their nuclear power plants, are currently enjoying a delightful holiday season.

In contrast, had nuclear weapons not been developed, World War III between the West and the Soviet bloc would almost certainly have broken out in the 1950s, likely followed by World War IV in the 1970s, World War V in the 1990s, and World War VI right about now.

In this respect the terror posed by thermonuclear weapons has thus far proven to be more of a blessing than a curse.

But if the unlimited power of fusion were to be put to work for purposes of creation rather than destruction, methods would have to be found to unleash it in a controlled way. Plasmas with temperatures of hundreds of millions of degrees centigrade would have to be generated and contained on a sustained basis without being allowed to contact any material significantly cooler. Stars can do this using their massive gravitational fields to hold in their hot plasma contents amidst the vacuum of space. But how could this be done on Earth?

The first person to grapple seriously with the challenge of engineering such a system was the Australian Peter Thonemann. While studying for his doctorate in nuclear physics at Oxford University in the immediate postwar period, Thonemann came up with the idea of employing the electromagnetic pinch effect.

An electric current going in a straight line creates a circular magnetic field around itself. The current, however, interacts with this field to generate a force that is perpendicular to both, pinching the current inward. Put another way, in the cylindrical coordinate system frequently employed by physicists, where Z is the direction along the cylinder, θ is the direction around the cylinder, and R is the direction of the radius of the cylinder, a current in the Z direction produces a magnetic field in the θ direction, and they interact to produce a force on the current carrier in the negative R (i.e. inward) direction. The power of this "pinch effect" had been demonstrated dramatically early in the 20th Century when lighting struck the chimney of the Hartley Vale Kerosene Refinery near Lithgow, New South Wales, Australia, causing a section of copper tubing to crush itself. Plasmas can be even more electrically conductive than copper, and when they carry a current, they heat up just like a wire, becoming superhot. But when compressed by the pinch effect their temperature is multiplied along with their pressure in accord with the gas laws of thermodynamics. Thonemann reasoned that such a "Z pinch" could be used to ignite fusion.[11]

Thonemann wrote up his ideas and sent them to Frederick Lindemann (aka Lord Cherwell). Lindemann had been Winston Churchill's wartime science advisor, and by 1946 was serving as the director of Oxford's Clarendon Lab. Lindemann responded by asking Thonemann to deliver a seminar to an audience of Clarendon's top physicists. Thonemann accepted the invitation and did a bang-up job. The boffins were onboard. Thus began the world's first thermonuclear program.

Thonemann's Z-pinch concept faced many challenges. The plasma needed to be contained in a vacuum vessel to keep air out. But this could not be built in simple cylindrical geometry because when drawn by an electrical field in the "Z" linear directions, the plasma's electrons and atomic nuclei (called "ions") would swiftly accelerate in opposite directions along the length of the vessel and hit its ends, extinguishing the plasma. So instead of a cylinder, the system needed to be built as a torus. In such a system a circular electric field created by electromagnetic induction could be used to make the electrons and ions race around the toroidal track, heating up the plasma without ever being driven into a reactor end.

That much was clear right from the start. Unfortunately, things were more complicated than that. Plasmas tend to be pushed by magnetic fields from regions of strong field to regions of weak field. But an electrical current moving in a circle will always create a stronger magnetic field on the side of its inside track than its outside track. This will create a force driving the plasma outward, causing the plasma current in a toroidal Z-pinch to move outward and hit the torus outer wall.

The British team was able to fight this tendency by placing conductors around the outside of the torus which created counter currents to repel the plasma's outward motion. But there were still more problems to deal with. Any small local bend that occurred within the current would also generate a stronger magnetic field on its inside track, generating a force that would make the bend grow bigger, until the kink became so sharp that the plasma current broke. If the current path should become constrained at any point, the magnetic field would become more intense at the constriction, which would drive the constriction even tighter until the current path was cut off altogether. These "kink" and "sausage" instabilities, illustrated

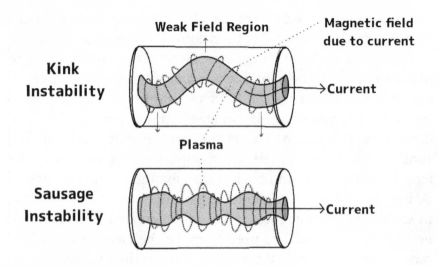

**Fig. 12.4.** The sausage and kink instabilities. Any disturbance that causes a local increase in the magnetic field of a Z-pinch current automatically magnifies itself until the current path is broken.

in Fig. 12.4, plague the Z-pinch concept, causing constant disruptions in operation.

The British found they could counter the kink and sausage instabilities by winding coils around the torus, creating a magnetic field that ran parallel to the ring current, thereby providing it with a sort of electromagnetic backbone. Proceeding in top secret fashion, they measured neutron emissions from the deuterium plasmas contained within their machines. This indicated that thermonuclear reactions were occurring. Greatly encouraged by such exciting results, they rapidly scaled up their devices, ultimately producing a large toroidal Z-pinch called Zeta by 1957.

In 1946, the US congress passed the McMahon Act, reneging on America's agreement to pursue nuclear technology development together with its British and Canadian Manhattan Project partners. In retaliation, the British kept their controlled fusion research secret from the Yanks. As a result, America's controlled fusion did not begin until 1952, when Princeton astronomer Lyman Spitzer, enjoying a ski vacation in Colorado, was surprised to read a newspaper article reporting a claim by Argentine dictator Juan Peron that his country had created a functioning fusion reactor.

Peron's claim was false, but it got Spitzer thinking. How could a thermonuclear fusion plasma be confined?

Since Spitzer was not Australian, the pinch-effect-driven destruction of the chimney of the Hartley Vale Kerosene Refinery was not at the top of his mind. Instead, he focused on making use of a different electromagnetic phenomenon.

When charged particles move in a direction perpendicular to a magnetic field a force is generated that is perpendicular to both. There is no force generated, however, if the particle is moving parallel to the magnetic field. The combination of these two effects makes charged particles spiral around magnetic field lines, moving along their length like railroad cars tied to a track.

In a simple solenoid, the magnetic field lines run parallel to the tube. Being composed of charged particles, a plasma inserted in such a system is restrained from expanding outward to hit the tube walls, but it is free to run out either end. However, if the solenoid were bent into a torus, then like an electric train set on a circular railroad (which were all the rage in the 1950s), all the plasma particles could do is run around the magnetic field line track.

The only problem was that, once again, in any toroidal arrangement, the magnetic field is stronger on the inside track of the toroid than it is on the outer track, and since plasmas like to move from regions of strong magnetic field to regions of weak field, this would drive the plasma outward to hit the toroid outer wall.

To solve this problem, Spitzer proposed that instead of using a simple toroid, his magnetic field track be laid down

**Fig. 12.5.** (Left) Lyman Spitzer with his first Model A Figure-Eight Stellarator. (Right) The larger Model B Stellarator on display at the Atoms for Peace conference, 1958. Credits: US AEC.

in the form of a figure-eight. As a result of this clever topology, the same magnetic field line that was on the inside track on one side of the apparatus would be on the outside track on the other. Plasma particles moving quickly along such a line would experience equal time on the inside and outside tracks, and thus feel no net outward or inward force. Spitzer called his invention a "stellarator," and moved quickly to build a small tabletop demonstration unit at his Princeton lab. The unit did not contain plasma all that well, as being so small, it leaked too much heat. But it showed promise. So, with AEC funding, but (unwillingly) under AEC secrecy requirements, Spitzer initiated a program to successively scale up stellarators to try to attain the temperatures necessary to sustain thermonuclear fusion. A mountain climber himself, Spitzer dubbed his reach for the stars Project Matterhorn.

Other American fusion initiatives soon followed. Working at the Livermore Lab, in 1952 Richard Post developed the concept of the magnetic mirror. Plasmas are pushed from regions of strong magnetic field. So instead of building complex magnetic toroids or figure eights, why not use a simple solenoid field in a cylindrical vessel, and stop plasma from leaking out the ends by plugging them with end magnetic fields stronger than the field in the middle.

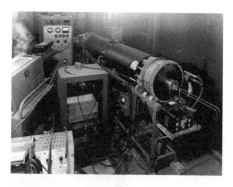

Post's team at Livermore built such a machine, and it was shown that the plasma did indeed leak out it ends much slower than it would have had the machine been a simple straight solenoid (a configuration also known as a "theta pinch") without

**Fig. 12.6** The Magnetic Mirror Q Machine at Livermore Lab, 1955. Credit: US AEC.

mirror fields at its ends. But it still leaked. Moreover, while the magnetic field lines in the linear section of the machine were straight, enabling stable confinement there, they bent inward towards the ends, with the concave side of the bend facing the plasma. This is what is known as "bad curvature," because just as in the field of a simple torus (which also has bad curvature on its side facing outward) it creates a situation where the field becomes weaker in the direction leading from the plasma to the vessel wall. So, since plasmas like to move from regions of strong field to regions of weak field, this creates an instability causing the plasma at the mirror ends to break out of conferment in the radial direction. To overcome this instability the Livermore team invented intricate coils for the mirror sections of their machine that could increase the field strength locally with good curvature. The solution was to design the mirror coils not as simple solenoid but in a manner similar to the stitchwork on a baseball. Using arrangements of such "baseball coils" or even more elaborate "Ying yang coils" the simple mirror's bad curvature instability could be suppressed.

But even more advanced mirrors leaked plasma out the ends, and every hot plasma particle that leaked took energy out of the system with it. The stellarator was also found to be leaky, due to "micro instabilities" caused by local collective plasma electromagnetic effects. Unlike the Z-pinch, both mirrors and stellarators lacked a powerful current internal current. So instead of being heated internally by electrical current resistance, a simple process known as "ohmic heating", they had to be heated using more complex techniques such as microwaves and charged particle beams. As a result, it was hard to get the plasmas in these machines to temperatures over 1 million centigrade, a far cry from the 100 million plus necessary for controlled fusion.

Frustrated with these problems, Jim Tuck, a Los Alamos scientist who had worked with Thonemann in the late 40s and so was aware of his ideas concerning fusion (if not his more recent secret efforts to implement them) started a toroidal Z-pinch program of his own. With due (but rare) humility of someone armed with 1950s science taking on the challenge of controlled fusion, the puckish Tuck called his machine the "Perhapsatron."

Thus, by the mid-1950s, there were four major fusion initiatives proceeding in the West, with a wall of secrecy dividing the three young American efforts from the older British program. The Americans were making slow progress as they grappled with the numerous instabilities caused by the collective effects of magnetically confined plasmas. The British however, were excited, as their larger Zeta machine was generating plenty of neutrons, indicating the presence of thermonuclear reactions in its plasma.

Then, on October 4, 1957, the Soviets fired a shot heard round the world. As little Sputnik circled the Earth, its beep-beep-beep radio beacon sounded louder than a five-alarm fire wake up call to the science establishments of the Western powers. The Reds were ahead! What did we have to match them?

The Brits thought they had something. Zeta! The power that lights the stars, caught in a bottle! Take that Ivan! They decided to go public. After a brief consultation with the Americans, they moved to declassify their fusion work.

On January 24, 1958, Cockcroft (the very same Cockcroft who as a youngster in 1932 had first generated fusion in an accelerator, but now the Nobel Prize winning Sir John Cockcroft and director of the Harwell Lab housing the Zeta) held a news conference attended by 400 reporters from around the world. He announced that Zeta had reached a temperature of 5 million degrees centigrade, generating thermonuclear reactions with 90 percent certainty. That night on television he expanded further. "To Britain this discovery is greater than the Russian Sputnik," he said. Zeta's temperature would soon be increased from 5 million to 25 million degrees. Britain would have a working fusion reactor within twenty years.

The press response was immediate. "THE MIGHTY ZETA," blared the *Daily Mail* the next day. "LIMITLESS FUEL FOR MILLIONS OF YEARS."

"ZETA SPELLS POWER EVERLASTING," trumpeted *The News Chronicle*. "Britain last night became – officially – the first country to prove that the H-bomb can be tamed to feed power hungry nations. Harwell unveiled Zeta, its man-made sun, to show we lead the world in H-Power unlimited."[12]

Forewarned about the Zeta announcement, AEC Director Lewis Strauss arranged for the Los Alamos Z-pinch group to reveal their results the same day. *The New York Times* covered both announcements in the same article, concluding that while the Los Alamos results were weaker than Zeta's, the US and Britain were "neck and neck" in the race for fusion power.

While the public may have been overjoyed by the news of the Z-pinch breakthrough, the American stellarator and magnetic mirror fusion groups were not. They had been upstaged. Examining the data behind the headlines, they soon found ground for criticism. When *Nature* published its article reporting the Zeta and Los Alamos results, the issue also carried a polite rebuttal by the gentlemanly Lyman Spitzer. The results were anomalous he said, as the reported neutron production rates were much higher than those theory predicted for a plasma at that temperature. Furthermore the reported plasma temperature was much higher than it could really be. "Some other mechanism [than fusion] would appear to be involved."[13]

Younger physicists working with the mirror group at Livermore were less polite. "It was all bunk," said Harold Furth, a talented grad student who would, years later, become the director of the Spitzer's Princeton Plasma Physics Lab.

In fact, the critics were right. The Zeta plasma had not been heated to thermonuclear temperatures. What had occurred was a phenomenon later identified as "runaway electrons." In common electrical conductors, like copper, electrical resistivity increases with temperature. In plasmas it is just the opposite. The hotter a plasma gets, the less resistive it is, with conductivity increasing eightfold for every fourfold increase in temperature. Furthermore, if a minority group within the plasma gets hotter, its resistivity drops more than the rest. So in a toroidal Z-pinch, where there is an electrical field accelerating electrons around the torus, the faster an electron goes, (i.e. the hotter it is), the easier it becomes to accelerate, until it becomes a runaway electron, moving at relativistic velocities. On the rare occasions that one of these nevertheless does crash into a nucleus, it can impart a great deal of energy. This will endow the impacted nuclei with the power to set off a fusion reaction, despite the fact that the plasma as a whole is much too cold to sustain fusion.

Aluminum can burn – if it is hot enough. But when sparks hit cold aluminum, there is no chance it will ignite. Runaway electrons in a cold plasma are like sparks hitting cold aluminum. The phenomenon may create a measurable bit of light, but it's not the same thing as a fire. But that is all Zeta had actually accomplished. It had not demonstrated *thermonuclear* fusion – i.e. reactions sustained by the plasma's own temperature – and those jealous of its claims lost no time in making that clear. In June 1958, the British issued an embarrassed retraction.

There was one group of people who were not surprised by either the Zeta announcement or its eventual retraction. This was the Soviet thermonuclear fusion group working at the Kurchatov Institute in Moscow. While the British and Americans cloaked their work from each other with veils of secrecy, the Soviets, through their superb espionage networks, were fully apprised of the efforts and achievements of both. In fact, in 1956, Kurchatov had travelled to the Harwell himself, and given a surprise lecture on the problems of controlled thermonuclear fusion. Without revealing his knowledge of the British program, he had generously warned his audience of the issue of runaway neutrons, and how they might delude naïve scientists into believing they had achieved fusion. Unfortunately for their reputation, the British team had not taken his advice to heart.

With the advantage of superior knowledge and freedom from any vested interest in the Western concepts allowing them to view the putative results of the Z-pinches, mirrors, and stellarators with a cold eye, the Soviets were able to develop and focus on a better idea. This was the tokamak.

The theoretical basis for the tokamak (a Russian acronym combining the words for toroidal chamber magnetic) was laid down in 1952 by Kurchatov Institute scientists Sakharov and Tamm. But it was the brilliant experimental physicist Lev Artsimovich who led the arduous effort to transform the concept from promising theory into successful reality. For this reason Artsimovich is justly known as "the father of the tokamak."

The tokamak resembles the toroidal Z-pinch in that it includes a toroidal vacuum chamber containing plasma with an induced electrical current running around the circuit. This current heats the plasma and also produces a "poloidal" magnetic field encircling it. But the tokamak also includes a set of magnets around the toroid, enclosing it like a circular solenoid, thereby creating a strong "toroidal" magnetic field running around the toroidal racetrack. The later versions of the toroidal Z-pinch also contained some such magnets, but in the tokamak they were much stronger, providing the dominant confinement field. Also, the circular Z-pinch generally looked like a hula-hoop, with the major radii of their toroids much larger than the minor radii. In contrast, tokamaks were shaped more like fat doughnuts. If you divide the strength of the toroidal field by that of the poloidal field of a toroidal fusion device and compare that quotient to the aspect ratio of toroid (its major radius divided by its minor radius), you come up with a dimensionless number known as the "safety factor," abbreviated "q." In a circular Z-pinch the toroidal field is much less than the poloidal field and the aspect ratio is large, so q is much less than 1. But in a tokamak, the toroidal field is much bigger than the poloidal field, and the aspect ratio is small, so q is always greater than 1. As a result of its large safety factor, the tokamak is resistant to the kink and sausage instabilities that plague the circular Z-pinch.

The tokamak is also somewhat like the stellarator, in that it is essentially a solenoid field bent into a closed loop. However instead of twisting the torus into a figure eight to compensate for the outward force on the plasma generated by the inside track magnetic field being stronger than that of the outside track, the tokamak combines the effects of its toroidal and poloidal magnetic fields to make the magnetic field lines themselves spiral around the torus. This not only allows the system to be built in a simpler physical geometry, but also endows the tokamak with a powerful plasma current internal ohmic heating system that the stellarator lacks.

The Zeta affair may have been a public relations disaster for the British, but it had the positive effect of bringing all the

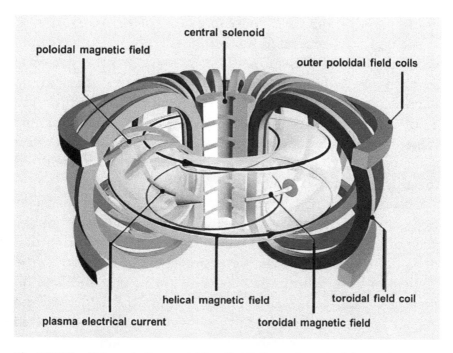

**Fig. 12.7** The Tokamak Concept (Credit: US Department of Energy)

world's fusion programs out into the open. At the 1958 Atoms for Peace conference in Geneva, all the players showed up, with the Americans making the biggest splash by bringing the actual Stellarator B fusion device and putting it on display in the exhibit hall. Being of a more practical mind, the Soviets enhanced the conference by bringing a set of attractive female translators. These quickly made fast friends with the all-male scientific contingents from the West, whose members had never before encountered so many vivacious young women with such intense interest in their theories and experiments. [14]

But while the Soviets may have been keen to learn everything they could about the West's fusion efforts by whatever means expedient, they also sought to assist them. This may seem paradoxical given the mendacious nature of the Soviet system, not to mention the Cold War context of the times, but it is so. While neutrons generated by fusion reactors can, in principle, be used to breed $^{238}U$ into $^{239}Pu$, there are many easier ways to do that – well-known and in practice by the 1950s. So the Soviets apparently (and correctly) deemed controlled fusion to be of no military significance. It could be of great economic value, however. So they wanted it developed. If the West's vast resources and scientific capabilities could be usefully mobilized towards that end, so much the better.

So, starting with Kurchatov's seminal 1956 lecture at Harwell, and continuing with Artsimovich's aggressive public promotion of the advantages of the tokamak (and brutal debunking of less productive approaches) in the 1960s, the Soviets did everything they could to try to put the West on the right track. This was not easy, because the leaders of the Western programs were wed to their own systems, and it suited them to dismiss the ever-better results being published by Artsimovich as Soviet propaganda.

Matters came to a head in 1968, when Artsimovich announced that his T-3 tokamak had reached plasma temperatures of 10 million Centigrade (or 1 keV), triple anything attained in the West. Met with the usual Western skepticism, Artsimovich did the unthinkable. He invited a British team to come to his lab, observe his experiments, and take their own measurements using their own instruments. The Brits accepted the invitation, and, using modern laser diagnostics unavailable in the USSR, confirmed the full truth of all of Artsimovich's claims.

The British report created a crisis in the US program. Led by Harold Furth, who had taken over the Princeton lab when Spitzer retired in 1965, the stellarator gang did not want to believe it. But there was no way the AEC could merely dismiss the news. In 1969 they appointed Fred Ribe (my future PhD advisor, but then leading the Los Alamos fusion program) to conduct a review. Ribe did so, and, faced with the hard data, came to the necessary

**Fig. 12.8** (left) the T-3 Tokamak. (right) Lev Artsimovich speaking to British scientists in Moscow, 1968.

conclusion: America needed to shift the central focus of its magnetic fusion program from stellarators to tokamaks. [15]

The Princeton group, still in (unrequited) love with its enlarged Model C stellarator, tried to fight the decision, but they were waging a losing battle. Tokamaks worked better than stellarators, that's all there was to it. When the Oak Ridge lab jumped into the debate and proposed that the nation's new flagship fusion device – a tokamak – be built in Tennessee, Furth realized he had to either adapt to the new reality or lose control of the game. He chose adaptation. If it's a tokamak you want, he said, Princeton can provide that too. In fact, the fastest way to make a sizable tokamak would be to reconfigure the Model C stellarator, which has almost all the parts needed. Just cut out the straight sections, put the two curved ends together to make a torus, and put a big iron core transformer through the center to induce the toroidal current, and presto, you've got yourself a tokamak!

As ad hoc as the idea sounded, it worked. Within six months the Model C Stellarator became the Symmetrical Tokamak, and under the successive directorships of Furth, Mel Gottlieb, and Dale Meade, Princeton became and remained America's flagship tokamak lab.

With the stellarator forces routed to near extinction (a few fanatics hung on in Germany) tokamak fever swept the West.

In 1971 The Texas Turbulent Tokamak went into operation at the University of Texas Austin, as did the ORMAK tokamak at Oak Ridge National Lab. These were followed by the Adiabatic Toroidal Compressor at Princeton in 1972, the: Tokamak de Fontenay aux Roses (TFR), near Paris, in 1973, Alcator A in at MIT, the Princeton Large Torus in 1975, Alcator C at MIT in 1978, TEXTOR in Julich, Germany 1978, TEXT at UT Austin in 1980, TFTR at Princeton in 1982, Novillo Tokamak, at the Instituto Nacional de Investigaciones Nucleares, in Mexico City, in 1983, the Joint European Torus (JET) at Culham, UK, in 1983, JT-60 in Japan in 1985, DIII-D at general Atomics in San Diego in 1986, Tokamak de Varrennes in and STOR-M at the University of Saskatchewan in Canada in 1987,1988: Tore Supra, at the CEA, Cadarache, France, in 1988, and Aditya, at Institute for Plasma Research (IPR) in India, in 1989. Meanwhile, the Soviets continued their program, advancing from the seminal T-3 through a series of ever larger machines including T-4 in 1969, T-10 in 1975, and the superconducting T-15 in 1988.

As a result of the tokamak fever, programs supporting alternative approaches to fusion, including not only stellarators, mirrors, toroidal Z-pinches, but novel ideas taking advantage of the self-organizing properties of plasmas, such as spheromaks and field reversed configurations were starved for funds. The last major old guard competitor to the tokamak was the magnetic mirror. But after spending over $200 million to complete the MFTF-B flagship magnetic mirror at Livermore, in 1985 the Department of Energy scandalously cancelled the program before the machine could even be turned on.

But if the fusion program was dangerously narrowed to one major concept, that approach — the tokamak — advanced vigorously.

The MIT Alcator machines were real workhorses. Their results provided a generally accepted scaling for the confinement time possible in a magnetic fusion device. According to this law, the average time a particle could be contained in a tokamak grows in proportion to the plasma density times the square of toroid minor radius.

Alcator scaling has decisive implications for tokamaks. The longer the confinement time, the less heat leaks out of the system, making it possible to get hotter. To obtain longer times, it is most critical that the tokamak be big. It is also important that its plasma be dense, which means that the pressure exerted by the tokamak's magnetic field must be strong enough to hold a lot of plasma pressure. Magnetic pressure increases in proportion to the square of the magnetic field strength. In mathematical terms, if T is the temperature, $\tau$ is the confinement time, n is the plasma density, H is the heating power applied to the tokamak, B is the magnetic field strength, and a is the toroid minor radius, then:

$$\tau \sim na^2 \qquad \text{(12.8) Alcator scaling}$$

$$H \sim nT/\tau \qquad \text{(12.9) Heating power = heat loss}$$

Combining these two equations, we find:

$$T \sim Ha^2 \qquad \text{(12.10)}$$

Then putting 12.8 and 12.10 together, we find that the Lawson parameter triple product of density × temperature × confinement time rises according to:

$$n\tau T \sim Hn^2a^4 \qquad \text{(12.11)}$$

Furthermore, since at a given temperature, the plasma density n will be proportional to $B^2$, it turns out that:

$$n\tau T \sim HB^4 a^4 \qquad\qquad (12.12)$$

Equation (12.12) is an extremely strong relationship. It says that the Lawson triple product rises in proportion to the fourth power of *both* the magnetic field strength *and* the reactor size, which means that doubling both should multiply the triple product 256 times! While tokamaks of any size can contribute to our understanding of plasma physics, only large machines with powerful magnetic fields can possibly reach the plasma conditions required for thermonuclear ignition. The effect of this law can be seen clearly in Fig 12.3, which shows how as tokamak size and magnetic field strength were increased from the 1960s through the 1990s, the critical triple product rose by a factor of 10,000. In tandem with this, as shown in Fig. 12.9, the actual amounts of fusion power released in the experimental tokamaks grew a trillionfold.

The three largest tokamaks yet built are the Princeton TFTR, the European JET, and the Japanese JT-60. All three of these were commissioned in the 1980s and reached their maximum performance in the 1990s or early 2000s with Lawson triple products within a factor of four of that required for fusion ignition. Such an improvement in performance could have readily been achieved by modestly scaling up these devices to the next generation of tokamaks. This almost certainly would have happened had the Olympic-spirted tokamak competition between the US, Europe, the Japanese, and Soviets continued beyond the 1980s.

However, this victorious march towards controlled fusion was halted dead in its tracks when the bureaucrats controlling the major fusion programs got together in the mid-1980s and decided that such competition was wasteful and stressful. It would be better, they reasoned, that instead of competing with each other to build more powerful tokamaks, we all got together

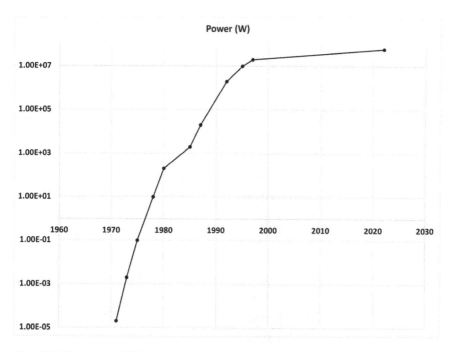

**Fig. 12.9** Between 1970 and 1997, the fusion power produced in experimental tokamaks grew a trillionfold. As no major new machines were built after the 1990s, however, progress in power level stopped. The JET team did continue to work to extend their burn time, however, producing a record 60 MJ by running JET for 5 seconds continuously at 12 MW output in 2022. Adapted from data in Reference 16.

on a single big machine, called the International Fusion Test Reactor, or ITER.

It was thus decided to abort the next generation of American tokamaks. Ribe, Meade, UCLA's Robert Conn, Fusion Power Associates leader Stephen Dean, and all the other veteran leaders of the US fusion program protested this course vigorously, but to no avail. Similar catastrophic decisions were made in Europe, Japan, and the Soviet Union.[17]

## THE ENTREPRENEURIAL FUSION REVOLUTION

So, after achieving considerable progress from the 1950s through the 1990s, for the past quarter century the world's government backed fusion power research programs have been mired in nearly complete stagnation. Is the situation hopeless?

No, it is not. The national fusion programs progressed well during the Cold War because of spirited international competition. They have stopped moving forward since late 1980s because the decision to consolidate them all into a single global project, the International Thermonuclear Experimental Reactor (ITER), removed all stimulus for action. Indeed, it took nearly a quarter century for the bureaucrats in charge of ITER to manage to reach a consensus in 2010 on where to put it, and it will take another quarter century before the machine even attempts to reach thermonuclear ignition in 2035.

This absurdly glacial rate of advance caused many people in and around the technical community to become cynical. "Fusion is the energy of the future, and always will be," became a common quip.

But a breakthrough has happened. Through its dramatic and rapid development of reusable launch rockets, Elon Musk's SpaceX company demonstrated that it is possible for a well-run lean and creative entrepreneurial organization to achieve things that were previously thought to require the efforts of governments of major power – and do so much faster and at much lower cost.[18] This hit observers of the fusion program like a bolt from the blue. Could it be that the seemingly insurmountable barriers to the achievement of controlled fusion – like the barriers to the attainment of cheap space launch – were not really technical, but institutional? Venturesome investors suddenly became interested.

I worked at Los Alamos in 1985 as a graduate student intern on a then-novel fusion concept called the spherical tokamak

(ST).[19] I can remember one lunch, when our group leader, Robert Krakowski, philosophically told the rest of us; "You know, when fusion power is finally developed, it won't be at a place like Los Alamos or Livermore. It will be by a couple of crackpots working in a garage."

We laughed at that then. Maybe Krakowski went a bit too far. But if not a couple of crackpots working in a garage, how about an entrepreneurial A-team of crack engineers working in a warehouse?

It's A-team time. A whole raft of innovative private fusion power startups is now getting funded. Here's a bit about some of them.

1.  **Tokamak Energy.** This Oxfordshire England venture, started in 2009 by former staffers from the Culham Laboratory Jonathan Carling, David Kingham, and Michael Graznevitch, has raised $50 million of mostly private money to try to develop the ST (the same concept that I worked on in the 1980s, which was too innovative for ITER to adopt) into a commercial reactor. In a magnetic

**Fig. 12.10** ITER under construction (left) Tokamak Energy's Spherical Tokamak (right) In 2022, Tokamak Energy's ST40 achieved 10 keV. (Credits ITER and Tokamak Energy.) © Tokamak Energy Ltd (2018)

confinement fusion reactor, the amount of power that can be generated rises in proportion to $b^2B^4$, where $\beta$ is the ratio of the plasma pressure to the magnetic pressure, and B is the magnetic field strength. An ordinary tokamak like ITER, can only achieve a $\beta$ of about 0.12, but an ST can achieve a $\beta$ of 0.4. As a result, an ST can produce the same amount of power as a regular tokamak in a machine less than 1/10th the size and cost.

2. **Commonwealth Fusion Systems.** Founded in 2018, this MIT-based venture has raised $75 million so far, including $50 million from the Italian oil company ENI and about $25 million from the Breakthrough Energy Ventures fund backed by Bill Gates, Jeff Bezos, Jack Ma, Mukesh Ambani, and Richard Branson. The root of the CFS design concept goes back to the 1980s, when the very creative maverick MIT physicist Bruno Coppi proposed achieving fusion in a small tokamak by the simple expedient of using ultra-strong magnetic fields. The magnetic field lines of a tokamak confine particles to follow them, spiraling around the chamber, with the radius of the spirals being inversely proportional to the strength of the magnetic field. Coppi reasoned that the relevant dimension of a tokamak was not its size per se, but the ratio of its size to the radius of the spiral, because it is this ratio that determines how long a particle will last before it hits the wall. Furthermore, as noted above, the higher the magnetic field strength, the faster the particle is likely to react. So if you want a particle to take part in a fusion reaction before it hits the wall (which would cool it too much for fusion), the key is just to go for broke with ultra-powerful magnets. But the problem is that the highest magnetic field it is practical to achieve with traditional low temperature superconducting magnets is about 6 Tesla, and Coppi needed 12 T. So, he designed an experimental machine called "Ignitor" using

copper magnets. This could not be a practical commercial reactor, because the resistive copper magnets would use too much power. Nevertheless, if it had been built, we probably would have achieved thermonuclear fusion ignition in the 1990s. But all of the US Department of Energy funds were committed to ITER, so Ignitor was never built. But starting around 2014, an MIT group led by Professor Dennis Whyte decided to pick up where Coppi had left off, improving on the Ignitor concept by making use of high temperature super conductor magnets, which require no electric power and can reach 12 T. As a result, with more than twice the magnetic field strength as ITER, the CFS reactor, known as SPARC (for Smallest Possible Affordable Robust Compact) fusion reactor, will achieve 1/5th the power hoped for by ITER in a reactor 1/65th the volume. Furthermore, CFS aims to do it by 2025, achieving in 7 years what ITER hopes to do in half a century.

3.  **Tri Alpha Energy (TAE).** Founded in 1998 by the late Dr. Norman Rostoker, southern California-based TAE has received over $800 million in investment from heavy hitters including Microsoft Co-founder Paul Allen, Goldman Sachs, Wellcome Trust, Silicon Valley's NEA, and Venrock. TAE's departure from orthodoxy is more radical than the above-mentioned startup in that they do not use a tokamak or toroidal chamber of any kind. Instead, TAE uses a simple cylinder chamber, with the required toroidal magnetic field induced in the plasma itself by having a linear magnetic field created by an outside solenoid suddenly reversed, causing it to curve around and connect to itself. This creates a kind of smoke ring current vortex in the plasma, or what is called in the fusion business a "field reversed configuration" or FRC. When I was in graduate school in the 1980s at the University of Washington, FRC's were all the rage, as they routinely achieve β values

**Fig. 12.11** Tri Alpha Energy has received over $800 million in private investment to develop the field reversed configuration into a practical fusion reactor. (Credit: TAE)

over 0.5. Moreover their simple cylindrical construction makes them potentially much more promising for creating low-cost commercial systems or fusion rocket drives than tokamaks. But by the 1980s, tokamaks had crowded out all funding within with US fusion budget, and shortly afterwards even the US tokamaks were starved from funds to feed ITER. FRCs were far too avant garde to even be considered by ITER. But private investors are much more daring than international bureaucrats, and TAE is pushing hard, with a goal of demonstrating net energy production by 2024.

4.   **Helion Energy.** Founded in 2005 by Prof. John Slough of the University of Washington, Helion uses two FRCs that are accelerated into a cylindrical reaction chamber

from opposite ends to collide in the middle where they are compressed by a solenoidal magnetic field to reaction conditions. Fusion reactions then heat the FRC plasma, causing it to expand back towards the chamber ends at high speed, with its energy being directly converted into electricity in the process. The cycle would then be repeated once per second to keep making power, or alternatively rocket thrust. In November 2021, Helion Energy announced the close of its $0.5 billion Series E, with an additional $1.7 billion of commitments tied to specific milestones. The round was led by Sam Altman, CEO of OpenAI and former president of Y Combinator. Existing investors, including co-founder of Facebook Dustin Moskovitz, Peter Thiel's Mithril Capital and the Capricorn Investment Group also participated in the round.

5.   **General Fusion.** Founded in Burnaby British Columbia by Dr. Michel Laberge and Michael Delage in 2002, GF has since received some $130 million in investment. The GF concept injects an FRC into a chamber containing a

**Fig. 12.12** In General Fusion's design, pistons drive a liquid wall inward to implode an FRC. (Credit: General Fusion)

rotating liquid metal wall, which is then driven inward by an array of pistons to compress the FRC to fusion conditions. This is a variant of the "imploding liner" concept that has a heritage going back to the AEC's 1972 Project LINUS. The theory behind it is complicated, but appears sound. GF hopes to show it will all work in the mid-2020s.

6. **Lockheed Martin.** In 2010, under the inspiration of Dr. Tom McGuire and Charles Chase, Lockheed Martin initiated its own "Compact Fusion Reactor" development program using internal funds. The CFR appear to be a linear cylindrical system, confined at the ends by increased magnetic fields (or "magnetic mirrors") but with an extra pair of superconducting magnetic coils operating inside the plasma chamber to make "cusps" that improve confinement. This creates a very attractive magnetic field configuration, but the engineering to make it work in an actual thermonuclear system seems quite challenging.

7. **EMCC.** In 1987, the late visionary Robert Bussard (of Bussard ramjet fame)[20] revived a 1950s concept originated by Philo Farnworth (the inventor of television) to use electrostatic fields, rather than magnetic fields, to confine a fusion plasma.[21] The idea works well enough that a very simple system can be used to generate a lot of fusion reactions, as demonstrated by neutron production, but all sorts of bells and whistles, including auxiliary magnetic fields are needed to get it close to generating net power. Bussard managed to get preliminary funding from the US Navy but now he is gone, and the rest of the team, led by Dr. Paul Sieck and Dr. Jaeyoung Park are seeking private funding. Any takers?

8. **Others.** In addition to the above, there are quite a few dark horses in the race. These include New Jersey based Lawrenceville Plasma Physics Fusion, led by Dr. Eric Lerner, who has produced interesting results using a concept called

the plasma focus; CT Fusion, a University of Washington based project founded by Dr. Tom Jarboe, Dr. Aaron Hossack, and Derek Sutherland, which is pursuing an FRC-like approach known as a "spheromak"; Applied Fusion Systems, founded in 2015 by Richard Dinan and Dr. James Lambert, who are trying their luck with an ST; Helicity Space, a company founded by Setthivone You, Marta Calvo, and Staphane Lintner, which is developing Dr. You's concept of taking advantage of the self-organizing plasma process that creates solar prominences to develop fusion reactors and fusion rockets; the Australian HB11 Energy company, which is applying Heinrich Hora's unique combined laser-magnetic fusion concept to try to ignite the p-$^{11}$B reaction; and Hyper V, NumerEx and the Sandia Lab/University of Rochester-based MagLIF project, which are all attempting to develop variants of the imploding liner concept.

Fusion is an unlimited source of energy, but there is an even greater power in the universe: human creativity. Fusion will give us wealth. Freedom will give us fusion.

## FOCUS SECTION: BUILD YOUR OWN FUSION REACTOR!

Now that you've heard all about how great fusion reactors are, you probably want to build one yourself. This is easy to do.

To save costs, I recommend that you forget about the magnetic confinement systems that the both the mainstream ITER program and most of the entrepreneurial companies are pursuing. Instead go with an electrostatic system, along the lines of that dreamed up by the great Philo Farnsworth. (He invented television too, so he probably knew what he was doing.) All you need to do to make one of these is create two spherical metal screen cages, one about half the diameter of the other, put the small one inside the big one, and then put the concentric assembly inside a vacuum chamber. Next, pump the vacuum chamber down to get rid of all the air, and put in a whiff of deuterium gas. (Believe it or not, you can order some out of the Aldrich catalog. Or just go diving in Lake Tinn.) Now **TAKING ALL REQUIRED SAFETY PRECAUTIONS** apply a large voltage difference between your two screen cages, (50,000 volts should do) with the outer screen charged positive and the inner screen charged negative. This will cause arcing to occur between the two screens, turning some of your gas into plasma, right in the middle of the powerful electric field between the screens. Separated from their normal electron pals that would otherwise make them neutrally charged and thus immune to influence by electric fields, the positively charged deuterium nuclei that you've created will hurtle inward, with energies of tens of thousands of volts. Some might hit the inner screen, but most will fly past it and instead collide with each other at super high velocity right at the center of the inner cage. Bang! You've caused nuclear fusion. It will look really cool too, so you should invite your friends over to watch it go. If they don't believe you've built a D-D fusion reactor, just tell them to bring over a neutron detector and measure the counts themselves.

While generating thousands of neutron counts, your little Farnsworth fusion reactor, (known as a "fusor") won't make enough reactions to be of practical use as a power generator, but you can still have a lot of fun with it. For example you can use it to bombard dimes with neutrons, turning their primary stable silver isotopes silver-107 and 109 into radioactive silver-108 and 110. Then you can give

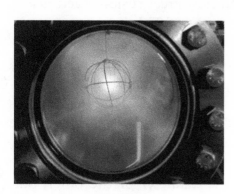

**Fig. 12.13** A Fusor in Action

radioactive dimes as presents to all your friends! (You can do this with quarters too, but dimes are cheaper.)

Enjoy!

NOTE: *The fusor experiment described here requires employment of very high voltage technology. Such equipment can be* **<u>extremely</u>** **<u>dangerous</u>. Do <u>not</u> attempt to do this experiment unless you are fully certified and experienced in the handling of high voltage technology.** *Also,* **the fusor construction instructions given here are <u>NOT</u> complete.** *Please consult all publicly available fusor literature before building your own. Also, be sure to take all necessary precautions to prevent any harm to yourself or others from the radiation that your fusor will produce. If you are a minor, be sure to get your parents' permission before building or operating a fusor as it may cause parts of your home to become radioactive, which could negatively impact its resale value and get you in trouble. Do not expose neighbors to fusor radiation without their knowledge and agreement. Be sure to comply with all applicable state and local regulations concerning the operation of home fusion reactors. If you take your fusor camping, be sure to only operate it in the officially fusor-approved areas of national parks. Fusor experimentalists proceed at their own risk, so* **please don't blame me if you electrocute yourself, irradiate yourself, or if the people in the apartment upstairs deal with you in a violent manner when they find out what you are doing.** *Thanks.*

# CHAPTER 13
# OPENING THE SPACE FRONTIER

**THE UNIVERSE IS** vast. It contains hundreds of billions of galaxies, each containing hundreds of billions of solar systems, each of which is incomparably larger than any one planet. Amidst such infinite prospects, it would be inconceivable folly for humanity to allow itself to be confined to one world. If we are to realize the potential of creating a truly grand human future, we must become spacefarers.

There are only two sources of energy that can support human activity in space. If you are close to a star, solar energy is available. But the vast majority of the universe is not close to any star. In all such regions, nuclear energy is the only option.

Within our own solar system, photovoltaic power is attractive for use in space out to about Earth's distance from the Sun. It faces limitations within that region for application on planetary surfaces, for example the Earth and the Moon, because of its periodic nature under such circumstances. These issues can be addressed at some cost with appropriate energy storage systems – with particular difficulties in the case of the Earth's Moon because of its four weeklong day-night cycle. However, as one moves outward from the Sun, available solar energy falls off in proportion to the distance squared. At Mars, the solar light flux is only 40 percent that prevailing on Earth, at Jupiter it is 4 percent, at Saturn 1 percent. In interstellar space it is essentially zero. If we wish to be able to do anything of consequence in most of the universe, we will need to use nuclear power.

## NUCLEAR ENERGY FOR SPACE APPLICATIONS

So far human activities in deep space have been limited to exploration. Nuclear power has played a critical role in enabling such activities. The *Viking* 1 and 2 Mars landers, the *Curiosity* and *Perseverance* Mars rovers, and nearly all probes to the outer solar system including *Pioneer Jupiter*, the *Voyager* 1 and 2 missions to Jupiter, Saturn, Uranus, and Neptune, the *Galileo* mission to Jupiter, the *Cassini/Huygens* mission to Saturn, and the *New Horizons* mission to Pluto, were all powered by radioisotope thermoelectric generators (RTGs, Fig. 13.1).

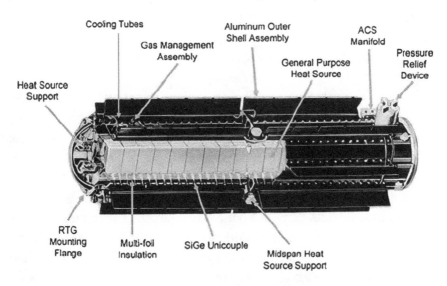

**Fig. 13.1** A Radioisotope Thermoelectric Generator (RTG) can generate electricity from heat with no moving parts. (Credit: NASA)

These units take the heat released by the 88-year half-life decay of Pu-238 bred for the purpose in nuclear reactors and convert it into electricity at about 5 percent efficiency using thermoelectric devices. Thus a typical (large size) RTG might use a 6-kilowatt radioisotope heat source to create 300 Watts of electricity to run a spacecraft. Such thermoelectric systems have thus far been favored by NASA because they have no moving parts.

However by employing a dynamic power conversion cycle, like a Brayton or Stirling engine, dynamic isotope power systems (DIPS) could increase the efficiency of radioisotope generators to 25 percent and get 1500 W of useful electricity from the same plutonium heat source that produces only 300 W in an RTG. Aware of these advantages, NASA has launched several programs over the years to develop Stirling cycle driven DIPS units. Unfortunately, after being allowed to spend a few hundred million dollars to bring DIPS to an advanced state of development, the programs have been repeatedly cancelled by bureaucratic or congressional caprice before being brought to flight status.

A much more impressive gain in capability could be obtained, however, by replacing radioisotope decay as the heat source with nuclear fission. In that case, 100 kWe units would be readily achievable even using thermoelectric conversion, with dynamic cycles enabling multi-megawatt systems. Such an advance would offer extraordinary benefits for planetary exploration.

The rate at which a spacecraft can transmit data is directly proportional to its power. So a Saturn probe, for example, seeking to study that planet and its extraordinarily interesting moons Titan and Enceladus, equipped with a 60 kWe nuclear reactor could send back 200 times the data as one dependent on a 300 W RTG. Furthermore, it would have plenty of power to do active sensing, for example probing the surface of Titan through its cloud cover and the oceans of Enceladus through their ice cover with high powered radar. This would enable discoveries that would be simply impossible without such capabilities. In addition, the probe could use its powerful reactor to enable extensive maneuvers using highly efficient electric propulsion. This would increase the spacecraft's useful payload, multiply its targets, and extend its duration, all of which would enormously increase the mission's science return.

**Fig. 13.2** The SNAP-10A Reactor Being Ground Tested in 1964. (Credit: US AEC)

The first space nuclear reactor to fly was the SNAP 10A, launched by the United States in 1965. Generating some 500 W of electrical power. The SNAP 10A operated for some 43 days, demonstrating both space nuclear power and electrical propulsion before the spacecraft failed for non-nuclear reasons. It is still in orbit.[1]

The initiative in space nuclear power then shifted to the Soviet Union, which flew some 33 nuclear powered Radar Ocean Reconnaissance Satellites (RORSats) between 1967 and 1988. Designed to scan the oceans for NATO ships, the RORSats needed to fly in low orbits because the strength of a radar return signal diminishes in proportion to the inverse fourth power of the distance between the transmitter and its target. Since in such low orbits, solar arrays would have created too much atmospheric drag, the RORSats had to be nuclear. Of the

**Fig. 13.3** A TOPAZ Reactor.
(Credit: NASA)

33 RORSats flown, 31 used 2000 W BES reactors, while the final two were powered by TOPAZ reactors employing 10% efficient thermionic conversion to generate 10 kilowatts of power. [2]

The largest space nuclear power system flown to date, TOPAZ reactors had a mass of 1060 kg each and employed liquid metal NaK coolant to transfer heat from its 96% enriched $^{235}UO_2$ reactor fuel elements to its thermionic electric conversion system. Zirconium hydride in the core moderated the neutron spectrum, allowing the TOPAZ to operate as a thermal reactor. Reactor control was achieved using a system of twelve rotating drums surrounding the reactor containing neutron-reflective beryllium oxide. One of the TOPAZ units operated successfully for six months in space, the other for 12 months.[3]

Following the fall of the Soviet Union, the dead-broke Russian Federation sold (all?) six of its remaining TOPAZ reactors to the American Strategic Defense Initiative Organization (SDIO), which hoped to develop a similar system for its own needs. However, when Bill Clinton was elected president in 1992, his fanatically antinuclear Vice President Al Gore took control of America's space efforts. As a result, both the SDIO's thermoelectric reactor program and NASA's concurrent SP-100 program aimed at developing a 100-kilowatt thermoelectric space nuclear reactor were immediately cancelled.

(Gore's effort to shut down America's nuclear space program was truly vindictive, and potentially illegal. In 1995, while employed by Lockheed Martin, I won the Outstanding Paper

Award from the Symposium on Space Nuclear Power in Albuquerque, by presenting a very powerful paper detailing the advantages of space nuclear power. In response, Lockheed Martin's top management received a phone call from Gore's office, instructing them to shut me up. The resulting chill on my publications was one of the reasons why I decided to leave Lockheed Martin and start my own company, Pioneer Astronautics. In 1999 I met Al Gore at the annual conference of the American Association for the Advancement of Science. When I politely handed him a copy of my book, *The Case for Mars*, he responded by throwing it on the ground. The Secret Service apologized to me for his behavior.)

Gore's destruction of the SP-100 and SDIO efforts left the United States without a space nuclear reactor program for twenty years. However in 2012, Los Alamos engineers David Poston and Patrick McClure began an underground effort to change the situation. Getting control of Flattop, a tiny nuclear reactor used for physics studies at the Nevada Test Facility, they hooked it up to a liquid potassium heat pipe and used its energy to drive a small Stirling engine. This system, known as DUFF (for Demonstration Using Flattop Fissions) then became the basis for a custom-built 1-kilowatt system known as KRUSTY (for Kilowatt Reactor Using Stirling Technology), which was successfully tested in 2018. This in turn led to the formal establishment of the joint DOE-NASA "Kilopower" program, which is currently ongoing. According to its charter, the goal of this program is to advance the KRUSTY system, which employed 93% enriched Uranium metal in fast reactor core, liquid metal heat pipes and Stirling cycle conversion, to develop operational space nuclear power reactors up to 10 kilowatts in capacity.[4] As the Kilopower program has transited from a buccaneering to an establishment effort, however, it has become more bureaucratized. This change in program culture has driven the original rebel-spirited inventors out and slowed progress considerably.

Nevertheless, it is to be hoped that Kilopower will eventually succeed regardless. As an outsider to the program, however, I must view it as a disappointment that nearly four decades after TOPAZ reactors flew in space, the highest ambition of the United States government's leading technical organizations is to duplicate their capability. Certainly 10 kWe nuclear reactors beat 300 W RTGs, so bring them on. But by this time, we should have been building space nuclear reactors with 100 kWe capacities, or more.

## NUCLEAR SPACE PROPULSION

Nuclear power can also be used to enable more capable types of space propulsion. Rockets work by the principle of conservation of momentum, according to which the faster a given mass of propellant is exhausted out the rear of a rocket, the more forward momentum it can impart to the spacecraft. This relationship is known as the "Specific impulse" of a rocket, abbreviated $I_{sp}$, are given in units of seconds. A rocket with a specific impulse of 450 s can make 1 pound of propellant deliver 1 pound of thrust for 450 s. (It would also have an exhaust velocity of 4410 m/s, which is easily calculated by multiplying 450 s times 9.8 m/s$^2$, an acceleration of 1 gravity). In fact, an $I_{sp}$ of 450 s is about the best you can possibly get with any chemical rocket. It's what you can get reacting hydrogen with oxygen to make water, *and it's all you can get* because the energy available to accelerate the water vapor exhaust is limited by the energy of the chemical reaction that produced it. However if you have a source of electricity in space, you can apply any amount of power to each unit of propellant, and get any specific impulse you want. The catch, however, is, that if you want the ratio of power to propellant mass to be very high, you need to keep the rate of flow of propellant low, which means that the thrust of the rocket will be low as well. The best $I_{sp}$ a Space Shuttle Main

Engine could do was 453 s, but it had a thrust of 2.23 million Newtons (500,000 pounds force, lbf.) Electric thrusters have been built and flown with an $I_{sp}$ of 5000 s. But with even with 1000 kWe of power to drive such a thruster (ten times that available on the International Space Station, and a hundred times that produced by the TOPAZ) the maximum thrust generated would be 28 Newtons (6.3 lbf.). So electric thrusters are like a car engine that gets 1000 miles to the gallon but takes an hour to go from zero to 60 mph. If you've got plenty of time to perform your maneuver, they are a great way to save propellant. But if you need pickup, they are useless.

In consequence electric thrusters are best for use for stationkeeping, to support slow maneuvers like puttering around the asteroid belt or gas giant moon systems, or to accelerate spacecraft over the course of a multi-year trajectories to the outer solar system. In the latter role, nuclear electric propulsion (NEP) shines, because solar power is unavailable. But they can't be used for quick trips to the Moon or Mars, let alone for ascent from a planetary surface.

There are, however, two other ways to make use of nuclear energy for space transportation. The first of these is known as a nuclear thermal rocket, or NTR.

In an NTR, a nuclear reactor is used not to produce electricity, but to directly heat a propellant, such as hydrogen. The propellant, which is stored onboard the spacecraft as a liquid, is blasted into high temperature gas as it flows through cooling channels in the reactor. Essentially, the system is a flying steam kettle. The amount of power the reactor can generate is not limited by an electrical conversion system, so any level of thrust is possible. However the specific impulse is limited by the maximum temperature the gas can be heated to, which in turn is limited by the maximum temperature the reactor materials can sustain. If the propellant is hydrogen or ammonia, uranium carbide embedded in graphite or carbide materials can be used, and sustain temperatures of about 3000 K. The maximum exhaust velocity of any thermal rocket is equal to the square root of twice the energy (or enthalpy, if you really want to get technical) per unit mass. For hydrogen propellant this works out to 9 km/s, while for ammonia, with its higher molecular weight (8.5 when dissociated, compared to hydrogen's 2), it is 5 km/s, resulting in specific impulses of 900 s and 500 s, respectively. Thus a hydrogen NTR can obtain a specific impulse twice that of the best possible chemical engines.

Rocket engineers will kill for a few extra seconds of specific impulse. Doubling the $I_{sp}$ of a high thrust engine from 450 s to 900s would be extraordinary. For this reason, in the 1960s the United States instituted a series of joint NASA/AEC programs dubbed first ROVER then NERVA to build and test NTRs. Under these programs twenty NTR engines with thrust levels running from 15,000 to 250,000 lbf were built and tested in the Nevada desert. They worked as predicted, reaching ~~from~~ WHAT

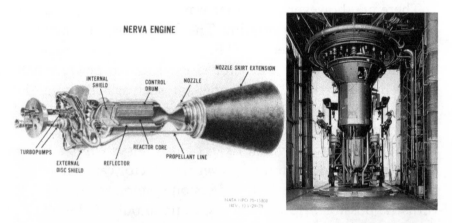

NERVA ENGINE

**Fig. 13.4** (Left) Diagram of NERVA NTR engine. Hydrogen flows through channels in the uranium carbide/graphite core, where it is heated to 3000 K then exhausted out a rocket nozzle. (right) The 55,000 lbf thrust NERVA XE being prepared for testing at Jackass Flats, Nevada, 1967. (Credits: NASA)

NASA calls Technology Readiness Level 6 (TRL-6), fully ground tested and ready for flight test.[5]

NASA planned to follow up the Apollo Moon missions with human missions to Mars in 1981, using 75,000 lbf thrust NERVA engines for propulsion. A ground test unit of such an engine, the NRX-A6, was fired for 62 minutes at Jackass Flats Nevada in 1968, achieving a specific impulse of 850 s and fully verifying the necessary operational capability.

However in 1970, the Nixon administration decided to curtail Apollo and cancel NASA's post Apollo plans, which included a permanent lunar base in the 1970s, human missions to Mars in the early 1980s, leading up to a permanent Mars base in the 1990s. (This decision was disaster of world historical dimensions. It was the equivalent of Ferdinand and Isabella meeting Columbus after his discovery of the New World and telling him they weren't interested. Had NASA been allowed to continue its post Apollo plans, the first children born on Mars would have graduated high school – on Mars – by now. Instead the funds were diverted to laying waste to Vietnam and Cambodia for a few more years, so that the Vietnam War, which was already lost, could be lost "with honor.")

Once the human Mars mission was cancelled, there was no longer a requirement for the NERVA program, so the Nixon administration sought to cancel it. The program however was defended strongly by Senate heavy hitters Alan Cranston (D-CA) and Clint Anderson (D-NM), who were able to continue it over Nixon's opposition. (The FDR-JFK Democrats were strongly pronuclear.) By 1973, however, the Democrats had begun to embrace reactionary Malthusianism, and with the removal of their support the NERVA program had to give up the ghost.

There was an attempt to revive the NTR program under SDIO auspices during the Reagan-Bush years. This included exploration of a promising new particle bed reactor (PBR) concept which offered the potential to quadruple the thrust to weight ratio of NTR engines.[6] NERVA engines had a thrust to weight ratio of about 4, making them only viable as systems for in-space propulsion. The fuel in PBRs was divided into tiny pellets, which collectively had a much higher surface to volume ratio than was possible using the NERVA type approach of drilling channels through as solid core. As a result, PBRs could be much more compact and lightweight. However, with the election of the Clinton-Gore administration the PBR effort was swiftly shut down. The United States has not had a serious NTR program since.

There is, however, another way to use nuclear energy in space to enable highly efficient propulsion. That is to use nuclear energy on the surface of a planet to transform local materials into chemical propellants and use them in a chemical rocket engine.

The advantage of this approach over nuclear electric propulsion (NEP) or NTR was made clear to me in the 1980s by Kraft Ehricke, a surviving member of Werner von Braun's rocket team. At that time I was a strong advocate of NEP. But Ehricke pointed out to me that if we used nuclear power to separate oxygen out of lunar iron oxide, we could combine that with hydrogen brought from Earth (the existence of lunar ice deposits was unknown in the 1980s) to power a hydrogen-oxygen engine like the workhorse RL-10 (which he had developed.) The RL-10 burns oxygen with hydrogen with a mixture ratio of 6:1 by weight to produce 20,000 lbf of thrust with an $I_{sp}$ of 450 s. Nothing too remarkable about that. But if you only had to launch $1/7^{th}$ of the propellant from Earth, then the *effective specific impulse* of the system would not be 450 s, but 7 × 450 s = 3150 s, which is *as good as electric propulsion but at high thrust!* Moreover the spacecraft would only have to carry a lightweight RL-10 instead of a heavy nuclear reactor and its massive shield.

It was this line of thought that led me to develop my Mars Direct mission plan in 1990.[7] In Mars Direct, a modest Earth

Return Vehicle is sent to Mars in advance of the crew carrying 6 tons of liquid hydrogen. Using a rover to deploy a 100 kWe nuclear reactor on the Martian surface, the ERV makes use of the power to combine its hydrogen with 102 tons of $CO_2$ from the Martian atmosphere (Mars's air is 95% $CO_2$) to produce 108 tons of liquid methane/oxygen propellant. Used in a modified RL-10, methane/oxygen can produce a specific impulse of 380 s. But since the amount of propellant ultimately produced is eighteen times as great as the 6 tons of hydrogen transported from Earth, the effective specific impulse of the system is a phenomenal 18 × 380 = 6840 s. It is this terrific performance, once again delivered at high thrust using an inexpensive lightweight reliable engine that has been in widespread use since the 1960s, that enables the Mars Direct plans' strategy of direct crew flight to Mars in a hab module, followed after 18 months surface exploration with a direct flight from the Martian surface back to Earth, without any need for on-orbit assembly, advanced propulsion, orbital motherships, or orbital rendezvous of any kind in any phase of the mission. Not only that, but the power requirement for the nuclear system used in Mars Direct is only 100 kWe, far less than the 10,000 kWe systems required for human Mars missions using nuclear electric propulsion (NEP). This is because by using nuclear power to transform local materials into propellants, the power generated by the reactor can be integrated over a period of a year or more and stored as chemical fuel, then released within minutes when it is required. In contrast, the NEP system must be much larger, because it needs to produce its thrust power in real time. Even so, a 10,000 kWe NEP system could only produce about 60 pounds of thrust, less than one-tenth of one percent the 100,000 lbf of thrust generated by a Mars Direct ERV equipped with five RL-10s. This makes the space nuclear power system used in Mars Direct much lighter, and vastly less expensive to both develop and build than that needed to attempt the mission using NEP.

**Fig. 13.5** The Mars Direct plan. The conical Earth Return Vehicle is sent to Mars first and deploys a 100 kWe nuclear reactor. Power from the reactor is used to react 6 tons of hydrogen brought from Earth with 102 tons of $CO_2$ drawn from Mars to produce 108 tons of methane/oxygen bipropellant. Once this is done, the crew flies to Mars in the tuna-can shaped hab module and lands nearby. After 18 months on Mars, the crew then flies directly back to earth in the ERV. The hab is left behind on Mars, so that as missions proceed, more habs are added to the base, creating the basis for the first human settlement on a new world. (Credit: Robert Murray)

The 10,000 kWe nuclear power system used in a human Mars mission NEP system would probably have a mass, including its shield, power conditioning, reactor and engines, on the order of 200 tons, greatly outweighing the ~40 crewed habitation module it is pushing. In contrast, the 100 kWe reactor system used by Mars Direct would have a mass of about 4 tons, and the propulsion system of the ERV wouldn't need to push it at all. The NEP system thrust is too low for ascent from Mars. It can only be used for in space propulsion. The Mars Direct

THE

system can enable ~~to~~ ERV to both takeoff from the Red Planet and travel from Mars orbit right back to Earth.

It gets even better. Since Mars Direct was proposed in 1990, massive ice formations have been discovered on Mars as far south as 38 N, which is the same latitude as San Francisco on Earth. A Mars Direct mission going to such a location could get both $CO_2$ *and* water from the Red Planet, enabling it to produce its return methane/oxygen propellant with no chemical feedstock transported form Earth whatsoever. The effective specific impulse of such a system would therefore be infinite. This is why Elon Musk's SpaceX company has decided to use methane/oxygen engines to drive its Starship launch vehicle,[8] and adopted a Mars Direct derived plan of flying the mission directly to Mars, and then refueling there with locally produced propellant for a direct flight home. The Starship needs 600 tons of propellent to fly from Mars to Earth. That can be produced on Mars using the energy content of 600 grams of uranium fuel. Instead of shipping 600 tons of chemical propellant to Mars to enable a single return flight, send a 10-ton reactor packing enough energy to make *6 million* tons of propellant.

The lesson: Don't bring your propellant to Mars, bring its energy, in nuclear form, and make your propellant there. With nuclear power, interplanetary explorers can travel light and live off the land. That's the smart way to travel

## ENERGY FOR MARS SETTLEMENT

The lessons of the Mars Direct plan apply more broadly, and with even greater force, if we shift our focus from Mars exploration to settlement. Of the worlds currently within our reach, Mars possesses by far the richest assortment of raw materials for transformation into resources. All the elements we have on Earth can also be found on Mars, everything else you need to not only refuel exploration missions but build cities and civilizations on the Red Planet can also be produced there, provided you have energy. If you want ceramics, glass, plastics, food, fabrics, steel, aluminum, silicon, uranium, deuterium – you name it – you can make it on Mars if you have power.

We can transform Martian materials into unlimited resources, but only if we have energy.

On Earth power can be generated by combustion of biomass or fossil fuels, tapping waterfalls and wind, converting sunlight by photovoltaic or concentrated thermal means, accessing geothermal sources, or nuclear power. On Mars, combustion is not a net energy source, as both the fuel and the oxygen to burn it need to be made. Water and wind power are unavailable, and will remain so until the planet is terraformed. Solar energy can be used on Mars, and has been, but the flux is only 40 percent as strong as it is on Earth, and it is subject to months-long cutoff by dust storms, making it even less attractive for large scale use than it is here. Geothermal power is a possibility, but only in limited amounts in limited locations. So the energy basis for a Martian civilization will need to be nuclear power.

Uranium and thorium have been detected on Mars widely dispersed at about 1 ppm concentration, which is about the same as is typical on Earth. Concentrated uranium ores, typically 2%, but in a few cases up to 20% uranium, have been found on Earth, and there is no reason why they shouldn't exist on Mars as well. So in principle, nuclear fission reactors could be fueled from uranium and/or thorium mined on Mars. The problem, however, is that refining high concentrations of these materials out of mined ores is a highly complex and involved industrial process, made much harder if it is necessary to enrich fissile material content from 0.7% U-235 to 3% U-235 via isotope separation to create reactor grade fuel.

We could, of course, ship fission fuel from Earth to Mars. This is certainly how initial Mars exploration missions will be powered. But if we wish to settle Mars, power requirements will grow exponentially, making support of the power systems by fissile imports increasingly logistically burdensome. One way to mitigate that problem would be to employ breeder reactors, which can obtain nearly all of the energy potentially available in typical fission fuel, as compared to about 2 percent in non-breeder systems. In this application, systems like the molten-salt breeder reactor seem the most promising, as they enable ongoing reprocessing of the bred fuel with minimal supporting industrial infrastructure.

Deuterium, however, is about 5 times as common in Martian water as it is on Earth (833 ppm on Mars vs 166 ppm on Earth.) Because the atomic weight of deuterium is twice that of ordinary hydrogen, separating them is much easier than separating U-235 from U-238, which only differ by weight by 1.3%. Moreover no mining is required, as a Mars colony will constantly be using water, and the physical chemical life support system will need to electrolyze at least 1 kilogram of water per day for every colonist. Thus a Mars colony of 100,000 people will need to electrolyze at least 100 metric tons of water per day, within which there will be 11.1 tons of hydrogen, from which 18.5 kilograms of deuterium can be extracted. This could be used as fuel in a D-D fusion reactor, whose net reaction, including highly reactive side products T and $^3$He which will be consumed in-situ, is:

$$3D => {}^4He + p + n + 21.6 \text{ MeV} \qquad (13.1)$$

The primary energy cost of hydrogen distillation methods of deuterium production is electrolysis. In this case, the electrolysis of water for life support will require about 480 MWe-hrs, or 20 MWe all day long. The amount of deuterium, however, will be enough to produce 1850 GW-hours, roughly 3,850 times as much energy as it took to extract, or 1,500 times as much power if we assume the fusion energy is converted into electricity at 40% efficiency.

If uranium can be mined and refined on Mars, the presence of plentiful deuterium would facilitate the building of reactors of the CANDU type, which use heavy water to moderate their neutron flux, and thus can employ natural uranium, without isotope enrichment as their fuel. This would at least eliminate the need for the very difficult U-235/U-238 isotope separation from the fuel cycle. Alternatively, natural uranium or thorium could be put in a blanket surrounding a D-D fusion reactor, and making use of the reaction's neutron product, be bred into fissile Pu-239 and U-233, which could then enable uranium and thorium breeder reactors.

But the simplest cycle is simply D-D fusion, made very attractive by the abundant deuterium that will be produced as a byproduct of the operation of the Mars' settlement life support system. It is for this reason that if fusion is not developed on Earth, it certainly will be on Mars.

It may be noted that deuterium currently sells for about $4,000/kg, so that the above 18.5 kg of deuterium would be worth about $74,000 per day, or $27 million/year if exported to Earth. That's not all that much. It could be worth a lot more if exported to the asteroid belt, whose enormous resources of precious metals cannot be mined without abundant energy.

Buy cheap, sell dear.

The same principle holds true among the asteroids, or the moons of Saturn. Matter is everywhere. If you have energy, you can turn it into anything you need. Nuclear power provides an extremely lightweight way to transport energy from Earth. Moreover, the fuel for nuclear energy – particularly fusion power – can also be found in space. As noted, deuterium is five times as common on Mars as it is on Earth, and the exotic helium-3 isotope, which can enable fusion reactions without any neutron production, can be found on the Moon and in the atmospheres of the outer planets – but not Earth. These facts may well drive space settlers to lead in fusion power development, with unbounded benefits to terrestrial humanity in consequence.

## THE ROAD TO THE STARS

Nuclear power can also enable new types of rockets, which we need if we are going to make it not just to our neighbor planets, but to other stars. The nearest star, Alpha Centauri, is 4.3 light years away. To reach it in a human lifetime, we are going to need to attain speeds at least a few percent of the speed of light, which is 300,000 km/s. But chemical rocket engines cannot produce exhaust velocities greater than 5 km/s because the energy per unit mass, or enthalpy, of chemical fuels is limited to about 13 megajoules per kilogram (MJ/kg) by the laws of chemistry. Nuclear fission, on the other hand, offers fuels with an energy per unit mass of 82 million MJ/kg, more than six million times as great as the best possible chemical propellants. Now the maximum theoretical exhaust velocity of a rocket propellant is equal to the square root of twice the enthalpy; thus 5.1 km/s for chemicals, 12,800 km/s for nuclear fission. That's a lot better. A fission rocket could thus, in principle, generate an exhaust velocity of 4 percent the speed of light. Since a spacecraft can generally be designed to obtain a speed (actually velocity change, or "ΔV") equal to twice its exhaust velocity, a theoretically perfect fission drive could get us to 8 percent lightspeed. Since Alpha Centauri is 4.3 light years away, that would mean a one-way transit in 54 years. If half the ΔV is used to slow down at the destination, maximum speed would be 4 percent of light, and the transit time would be increased to 108 years.

There are a number of problems, however. One of them is being able to take advantage of all the energy available. Primitive nuclear propulsion systems, such as nuclear thermal rockets, do a very poor job of this. These use a solid nuclear reactor to heat a flowing gas, and the maximum exhaust velocities attained are only in the 9 km/s range–good by comparison with chemical rockets, but nowhere near the performance needed for interstellar missions. If the nuclear fuel is allowed to become gaseous, (a "gas core" nuclear thermal rocket–NASA did a fair amount of work on such systems in the 1960s),[9] exhaust velocities of 50 km/s could be achieved.[10,11] This would be excellent for interplanetary travel, but is still not in the interstellar class. If a nuclear reactor is used to generate electric power to drive an ion engine (nuclear electric propulsion or NEP), exhaust velocities of up to several hundred km/s could be obtained, if hydrogen is employed as propellant. But the systems required are very massive, the thrust (and thus rate of acceleration) they can produce is low, and the exhaust velocity is still not good enough for interstellar missions. What is needed is a way to turn the nuclear energy directly into thrust. One answer is so straightforward it has been known since 1945: Use atomic bombs.[12]

It's clear that if you detonate a series of atomic explosives right behind a spaceship you can push it along rather well. Of course, if you don't go about it correctly, you might also vaporize the spaceship, or blow it to pieces, or turn the crew to jelly with a hundred-thousand Gs of acceleration, or kill everyone on board with a lethal dose of gamma rays. As we say in the engineering business, "These concerns need to be addressed." So, you must do it correctly. But if you can, you've got yourself one hell of a propulsion system.

This was the idea behind Project Orion, a top-secret program funded by the U.S. Atomic Energy Commission that ran between 1957 and 1963. The original idea came from Los Alamos bomb designer Stanislaw Ulam, and the program drew the talents of such visionary weapon makers as Ted Taylor and Freeman Dyson. A diagram of one of the Orion designs is shown in Fig. 13.6. In it, a magazine filled with nuclear bombs is amidships. A series of bombs is fired aft down a long tube to emerge behind the "pusher plate," a very sturdy object backed-up by some heavy-duty shock absorbers. When the bomb goes off, the pusher plate shields the ship from the radiation and heat and takes the impact of the blast, which is then cushioned by the shock absorbers. Since the bombs are detonated one after another in rapid succession, the net effect would be that a fairly even force is felt by the ship and its payload and crew, who are positioned forward of the bomb magazine. The pusher plate scheme is much less efficient at converting explosive force to thrust than a conventional bell-shaped rocket nozzle, (perhaps only 25 percent compared to the 94 percent that is state of the art), but it has much more force to play with. So maybe the real effective exhaust velocity would only be about 1 percent the speed of light. That puts a bit of a crimp on our plans for fission-driven interstellar flight, but still, an exhaust velocity of 3,000 km/s in a high thrust rocket has got to be considered pretty good.

However, for better or for worse, the Orion project came to a screeching halt in 1963 when the Test Ban Treaty between the United States and the Soviet Union banned the stationing or detonation of nuclear weapons in outer space.

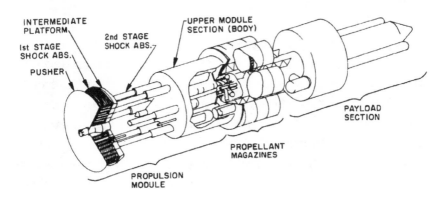

**Fig. 13.6** Orion Nuclear Bomb Driven Spacecraft. The bombs are dropped down a center tube and explode behind the pusher plate which absorbs the shock and shields the payload section. (Credit: Atomic Energy Commission)

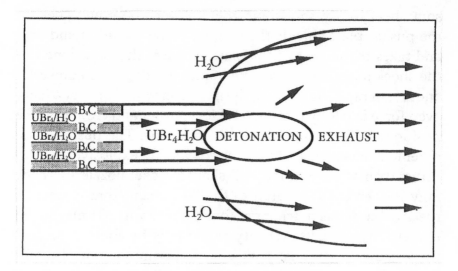

**Fig. 13.7** Nuclear Salt Water Rocket (Credit: Author)

The Test Ban Treaty will expire someday. But still, it seems like a good idea to avoid stationing ships in space filled with thousands of atomic bombs. I proposed a way around this problem in the early 1990s with a concept called a Nuclear Salt Water Rocket[13] (NSWR) shown in fig. 13.7.

In the NSWR, the fissionable material is dissolved in water as a salt, such as uranium bromide. This is stored in a bundle of tubes, separated from each other by solid material loaded with boron, which is a very strong neutron absorber and therefore cuts off any neutron traffic from one tube to another. Since each tube contains a subcritical mass of uranium, and the boron cuts off any neutron communication from one to another, the entire assembly is subcritical. However, when thrust is desired, valves are open simultaneously on all the tubes, and the salt water, which is under pressure, shoots out of all of them into a common plenum. When the moving column of uranium salt water reaches a certain length in the common plenum, a "prompt critical" chain reaction develops, and the water explodes into nuclear heated plasma. This then expands out a rocket nozzle that is shielded from the heat of the plasma flow by a magnetic field. In effect, a standing detonation similar to chemical combustion in a rocket chamber is set up, except that the enthalpy available is millions of times greater. The nozzle would be much more efficient than the Orion pusher plate, but because the uranium content of the propellant is "watered down," the exhaust velocity would also be decreased significantly below nuclear fission's theoretical maximum of 4 percent lightspeed, perhaps to about the same 1 percent achievable by a nuclear fission bomb driven Orion. But at least the need to buy a lot of extremely overpriced fission bombs would be eliminated.

With exhaust velocities of about 1 percent the speed of light, interstellar spaceships driven by such systems might be able to attain 2 percent light speed, allowing Alpha Centauri to be reached in about 215 years. Voyages with trip times on this order might be able to use rotating hibernations to allow a crew to reach the destination. Alternatively, there at least would be some chance that a multi-generation interstellar spaceship could reach its goal with its sense of purpose remaining intact.

However, in addition to offering only marginal performance for interstellar travel, such fission drives have another problem—fuel availability. The amount of fissionable uranium 233, 235 or plutonium 239 needed to fuel such systems would be enormous—perhaps several thousand tons to send a 1,000-ton (small for a slow, long-duration interstellar spaceship) payload on its way. That could get really expensive

We therefore turn our attention to a much better choice for interstellar spaceship propulsion, thermonuclear fusion.

## FUSION PROPULSION FOR INTERSTELLAR TRAVEL

High exhaust velocity is key to interstellar rocketry, and enthalpy is the key to exhaust velocity. Nuclear fission looks attractive at 82 million MJ/kg, but nuclear fusion is better. For example, if pure deuterium is used as fuel and burnt together with all intermediate fusion products (a series of reactions known as "catalyzed D-D fusion") 208 million MJ/kg of useful enthalpy is available for propulsion, plus 139 MJ/kg of energetic neutrons which, while useless for propulsion, can be used to produce on-board power. If a mixture of deuterium and helium-3 is used as fuel, the useful propellant enthalpy is a whopping 347 million MJ/kg. As a result, thermonuclear fusion using catalyzed D-D reactions has a maximum theoretical exhaust velocity of 20,400 km/s (6.8 percent of lightspeed, or 0.068 c) while a rocket using the D-$^3$He reaction could theoretically produce an exhaust velocity of 26,400 km/s, or 0.088 c.

Now we're talking starflight! With quadruple the enthalpy of nuclear fission and much more plentiful fuel, nuclear fusion holds the potential for a real interstellar spaceship propulsion system. As in the case of nuclear fission, fusion offers both pulsed explosions and steady-burn options for rocket propulsion, but in the case of fusion, both are more practical to implement.[14]

Fission bombs must be of a certain minimum size, because for a fission chain reaction to occur, a "critical mass" of fissile material must be assembled. Unless one chooses to simply waste energy by designing an inefficient explosive (a choice which is not a viable option for interstellar propulsion), this critical mass implies a minimum yield for a fission bomb of about 1,000 tons of dynamite.

Fusion is different. There is no critical mass for nuclear fusion, so in principle, fusion explosives could be made as small as desirable. Current military fusion explosives—H bombs—have very high yields, because they use a fission atomic bomb to suddenly compress and heat a large amount of fusion fuel to thermonuclear detonation conditions. If one wished to be crude, one could use such hydrogen bombs in an Orion-type propulsion system, with considerably higher performance and much cheaper fuel than the A-bomb driven version. However, with fusion there are other ways to achieve the required detonation effect on a much smaller scale.

This brings us to the subject of laser fusion. It is possible to use a set of high-power lasers to focus in on a very small pellet of fusion fuel, thereby heating, compressing, and detonating it. Preliminary experiments have proven the feasibility of such systems, and one, the National Ignition Facility or NIF, at the Livermore Lab in California, has demonstrated the ability to ignite thermonuclear explosions.

The achievement of ignition by NIF was announced with considerable fanfare by Energy Secretary Jennifer Granholm of December 12, 2022. According to Granholm, on December 5, 2022, NIF had focused an array of 192 lasers on a pea-sized pellet of deuterium-tritium fusion fuel and produced a thermonuclear reaction generating more energy than that used to drive it. The announcement was quick to set off headlines welcoming the advent of a vast new carbon-free energy resource for humankind, which were quickly followed by rebuttals from killjoys pointing out the vast distance between NIF and a practical commercial fusion reactor.

Many of the points raised by the killjoys were correct. While the NIF fusion fuel produced 3 megajoules (MJ – a MJ is about the energy needed to power a 100-Watt lightbulb for 167 minutes) of energy after being hit by a 2 MJ laser pulse, it took over 200 MJ of electricity to drive the lasers. So, from a practical point of view, no net energy was produced. Furthermore, the NIF, a football stadium sized facility built at the Lawrence Livermore National Lab (LLNL) at a cost of $3.5 billion, does not remotely resemble a power plant. Despite its immense capital cost it includes no mechanism for translating its fusion output into electricity. Furthermore, even if its electricity consumption could be reduced to zero, at the yield demonstrated it would need to be fired a thousand times a second to generate energy at the rate of a major urban nuclear or fossil fuel electric power station. The best NIF can actually manage is about one shot per day.

Yet if the critics made some valid points, they missed the main event. NIF didn't just achieve *breakeven* – getting more energy back that it put in – it achieved *ignition*. That is, they lit a thermonuclear fire in the lab. This has never been done.

To understand the significance of the NIF experiment, imagine that you are a Stone Age human, living in a society whose only source of fire is from lightning strikes. You observe that if you rub two sticks together they get warm. So you hit on the idea of rubbing them really fast and hard in order to try to light a fire artificially. After many tries and much effort, you manage to light a dry leaf on fire. The energy the burning leaf releases is much less than what you put in with your muscle power. But now you have a way to produce fire on-demand. That is what was accomplished at NIF.

The road to the NIF breakthrough was long and hard. Thermonuclear fusion reactions require temperatures far higher than that which any solid materials can withstand, so their ultra-hot fuel gas, known as plasma, needs to be confined without the use of solid containers. Stars can do it because they are surrounded by the vacuum of space, and can use their massive gravity to hold onto their fuel. Magnetic fusion devices, like the doughnut-shaped tokamaks discussed in Chapter 12, put tenuous plasmas in vacuum chambers and employ powerful magnetic fields to keep the plasmas from contacting the chamber walls. Inertial fusion devices, like NIF, attempt to avoid the necessity for either star-class gravity or massive magnets by heating the fusion fuel up so fast that it doesn't have time to go anywhere before it can react.

The concept of generating a massive explosion using inertial fusion was first hit on by Enrico Fermi in 1941. The Manhattan

Project ultimately focused on nuclear fission as a more practical way to create a weapon to win World War II. But Edward Teller became obsessed with the idea of using fusion to make a much more powerful explosive, and ultimately achieved success with the development of the hydrogen bomb in the early 1950s. However to obtain the power to heat and compress their fusion fuel fast enough to achieve ignition, H-bombs need to use fission bombs as triggers. Thus their minimum yield per detonation is that of their atom-bomb trigger, which is far too large to use to employ as a rapid-fire system for generating electric power. Furthermore, while the fusion reaction itself generates no radiative waste, the fission bombs used as H-bomb triggers produce plenty of fallout.

If inertial fusion was to be used to produce power, much smaller explosions needed to be generated, and a way to trigger then without atom bombs had to be found. In 1960, John Nuckolls proposed triggering them using lasers. Working at Livermore with Teller's protégé Lowell Wood and others, by 1972 he was able to publish a paper providing the math that showed it could really work.[15] Serious experimentation attempting laser fusion soon followed.[16]

In 1988 Nuckolls became Director of LLNL and advanced a proposal for a massive scale up of laser fusion research by building NIF. While the DOE's main drive for fusion reactors was focused on collaborating in the International Thermonuclear Experimental Reactor (ITER) tokamak project), Nuckolls argued that using NIF thermonuclear bomb physics could be researched without the need for actual underground bomb testing. This argument ultimately won the day with the Clinton Administration and the NIF program was begun. Conservative calculations indicated that it would be best to design NIF to be able to deliver 10 MJ laser pulses to assure ignition. But there wasn't the budget for that. Two MJ would have to do, making success chancy. The LLNL team took the bet, and NIF was completed in 2009. For thirteen years the team struggled to overcome all sorts of problems to make it work as they hoped. In December 2022 they finally pulled it off.

So now we have a way to produce fusion fire on demand. The knowledge we will gain from this capability will greatly benefit all other fusion efforts. This includes the inertial fusion program at the University of Rochester, which limited by funding to much smaller lasers than those employed by NIF, has developed a much more efficient approach to compressing and heating their fusion fuel targets. It also includes the whole raft of entrepreneurial magnetic fusion efforts mentioned in the previous chapter.

On December 2, 1942, Fermi's Chicago Pile-1 team achieved the first critical nuclear chain reaction. Following this success, Arthur Compton, who was there representing FDR's science advisory group, picked up the phone and called James Conant in Washington, saying "The Italian navigator has landed in the New World."

Nearly eighty years to the day later, the NIF team landed us in another.

An interstellar spaceship utilizing such a system for propulsion would eject a series of pellets with machine-gun rapidity into an aft region of diverging magnetic field. As each pellet entered the target zone, it would be zapped from all sides by an array of lasers. It would then detonate with the force of a few tons of dynamite, and the ultra-hot plasma produced would be directed away from the ship by a magnetic nozzle to produce rocket thrust.

Alternatively, using the knowledge generated by NIF, we may be able to learn how to implode and detonate fusion pellets using an appropriately shaped set of chemical explosives. If feasible, such chemically ignited fusion micro-bomblets would eliminate the need for a heavy laser system aboard ship.

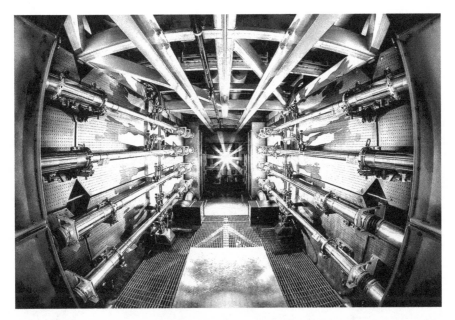

**Fig. 13.8** On December 5, 2022, the National Ignition Facility ignited a thermonuclear reaction in a deuterium-tritium pellet. (Credit: US DOE)

As a third alternative, one could implement fusion propulsion without bombs, lasers, or micro-bomblets, by using a large magnetic confinement chamber to contain a large volume of reacting thermonuclear fusion plasma. This is presumably a type of system that would be used to produce fusion power in the future, except in such a fusion drive most of the ultra-hot (tens of billions of degrees, or several megavolts) fusion products would be allowed to leak out of one-end of the reactor to produce thrust, while the rest would be used to heat the plasma to 500 million degrees (50 kilovolts) or so, which is the proper temperature for fusion reactors. Some of the lower temperature plasma would also leak out, but because of its lower energy, it could be decelerated by an electrostatic grid and used to produce electric power for the ship.

The magnetic nozzles used by fusion propulsion systems would not be as good as the 94 percent efficient bell nozzles used in chemical rocket engines but would be much more effective at channeling thrust than the 25 percent efficient pusher plates of the old Orion. Probably an efficiency of about 60 percent could be achieved. Assuming that to be the case, then a D-$^3$He fusion rocket should be able to attain an exhaust velocity of about 5 percent

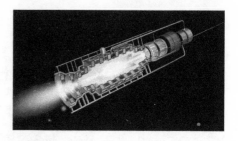

**Fig. 13.9** Magnetic Confinement Fusion Propulsion System. A magnetic field confines a reacting plasma in a magnetic mirror chamber, which is open to plasma escape at one end, thereby producing thrust. (Credit: NASA/ Princeton Satellite Systems.)

the speed of light. Since practical spacecraft can be designed to reach a speed about twice their engines' exhaust velocity, this implies that such fusion propulsion systems could make 10 percent lightspeed. Ignoring the small amount of extra time needed to accelerate, that means one-way to Alpha Centauri in 43 years, or 86 years, if we need to use the propulsion system to slow down.

Fusion power lights the stars.

*Ad astra, per astra.*

To the stars, through our stars.

# CHAPTER 14

# UPGRADING THE EARTH

**IN RECENT YEARS** it has been frequently claimed that humans are damaging the Earth. But if humans can damage the Earth, they can also improve the Earth. Furthermore, this is true of other species, and natural forces as well. The Earth has changed a great deal over the centuries and eons of its existence, and there is little basis for asserting that its condition at any particular point of time was the optimum state from which all possible changes represented deterioration.

How can we even know if we are improving or damaging the Earth? Well, if we consider the planets in our solar system, it is clear that some worlds are better than others. Mars is a cold dry desert, either lifeless, or at best, hosting a biosphere limited to subsurface microbes. Venus is pure hell, with a crushing cloudy and acidic atmosphere and surface temperatures hot enough to melt lead. While some overwrought aesthetes might argue that these planets are perfect in their current condition, anyone who tries settle on one of these worlds would surely disagree. Indeed, there is already an active discussion among those interested in launching new branches of human civilization on Mars of the possibility of "terraforming" the Red Planet, that is, making use of engineering techniques to make the Red

Planet more like Earth. Some critics have responded to these discussions by labeling such proposed projects as environmental crimes. But this is madness. If anyone was to attempt ᵗᵒ take the Earth as it is and transform it into something like Mars, that indeed would be a monstrous crime, and recognized as such by every sane person. That being the case, then transforming a dead planet like Mars into a wondrous living world like the Earth, filled with forests, meadows, and coral reefs, leaping dolphins, laughing children, great cities, universities, and used book stores, must be an action noble beyond reckoning. Were its accomplishment possible, anyone who would abort it would have to be numbered among the most despicable fiends in all of history.

Human beings, on average, create more than they destroy. This must be the case, or there would be nothing here. Furthermore, as our technology advances, our power of creation is growing exponentially. Humans will soon be able to travel to Mars, and being human, will seek to improve their new home. This being so, I believe that human settlers will eventually terraform the Red Planet. But long before we do that, we will upgrade the Earth. After all, we are here in force already.

The ability of humans to change planetary environments can be demonstrated by the effect we are already having on the Earth. Considering this necessarily brings us into a discussion of climate change, a scientific issue that unfortunately has been corrupted by political actors seeking to obscure, cherry pick, or exaggerate various aspects of the case for partisan purposes. Nevertheless, despite the fact that any objective discussion of this matter is sure to evoke outrage by militants hailing from both wings of the spectrum, I will try to discuss this important matter in as level a manner as possible.

First of all, global warming is quite real. Indeed, it is demonstrable that the Earth has been warming for the past 400 years. During Elizabethan times, the Thames River used to freeze over every winter, creating a place for "frost fairs" and other festivities. The last such fair was held in the mid-1600s, but as late as the mid-1800s, Charles Dickens could describe snowy winters in London, which no longer exist. During the American Civil War, Confederate soldiers stationed as far south as Georgia amused themselves by holding massive snowball fights pitting one regiment against another. By the early 20th Century, such climates were things of the past. Given the limitations of human industry over the period from 1600 to 1900, it's pretty clear that the warming that occurred during that span was natural, rather than anthropogenic. That said, in the 20th Century the rate has picked up, in a manner that is consistent with the predictions of the more conservative climate models for $CO_2$-driven global warming. On the basis of the available data, it would appear that coincident with a 50 percent rise of atmospheric $CO_2$ levels from 280 parts per million (ppm) to 420 ppm over the past 150 years (which is

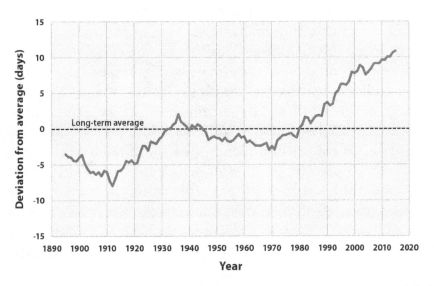

**Fig. 14.1** The length of the growing season in the United States has expanded markedly since 1895, a clear proof of climate change. (Credit: US EPA)

consistent with substantial human fossil fuel use), global temperatures have risen an average of about 1 C. This result may be doubted by some, because measuring an average global temperature to that degree of accuracy is a difficult task, with the results easily influenced by the choice of location for the thermometers. Nevertheless, it is clear that substantial warming has occurred because the average length of the growing season (i.e. the time between that last killing frost of the spring and the first one of the fall, an easy measurement to make) has expanded markedly over this period. For example, as shown by the EPA data presented in Fig. 14.1, the average length of the growing season in the USA has expanded by about 20 days since 1910, which is about as clear a demonstration of warming as you can get.[1]

Unfortunately, since this is a beneficial change, those attempting to make the case for a climate emergency never mention it, leaving their argument hanging on unconvincing claims of statistically averaged precision temperature measurements.

Global warming should also lead to greater net rainfall *on average*, and this indeed has also been measured, although the degree of change varies locally, with some regions actually experiencing drought despite an overall increase in the total.[2]

In addition to driving warming, the easily measured and clearly anthropogenic 50 percent increase in atmospheric $CO_2$ is also having a powerful direct effect on the biosphere, with rates of continental plant growth worldwide accelerated by about 20 percent. This result, which is consistent with the well-established theory of photosynthesis, is supported by innumerable lab studies, field studies, and most strikingly by satellite observations, as presented by the NASA data in Fig. 14.2, which shows a dramatic increase in the rate of plant growth over the past 36 years.

These findings provide grounds for dismissing some of the more strident warnings of climate activists. But let's not be

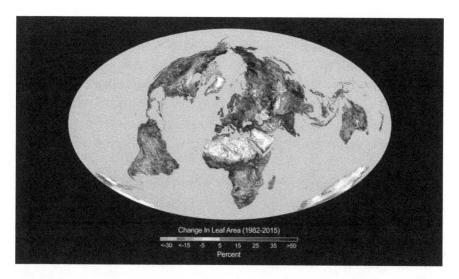

**Fig. 14.2** Leaf area has increased an average of about 20 percent world-wide over the past 36 years, a beneficial effect of rising global $CO_2$ levels. But what is happening in the oceans? (Credit: NASA).[3]

too hasty. While the moderate warming and significant $CO_2$ enrichment of the atmosphere we have experienced *thus far* has been, on the whole, rather beneficial, *unconstrained* temperature and $CO_2$ increases could be another matter altogether. Much larger increases are possible, as simply raising the rest of the world to the *current* US standard of living would require quintupling global energy production, with even more needed in reality due to population increase and the continued rise of living standards in the advanced sector. Furthermore, the happy picture presented in Fig. 14.2 only shows what is occurring on land. Most of the Earth is covered by oceans, and they show little evidence of enhanced biological productivity due to increased $CO_2$. On the contrary, significant damage to coral reefs appears to be occurring due to $CO_2$-driven ocean acidification (although conventional pollution and overfishing may also be to blame in some cases.)

So it would appear that massive anthropogenic $CO_2$ emissions are fertilizing the land while harming the oceans. This is happening because while $CO_2$ availability is a limiting factor for the growth of land plants, in most of the ocean the rate of growth of the phytoplankton that stand at the base of the food chain is controlled by the availability of trace elements, such as iron, phosphorus, and nitrates. This is why well over 90 percent of the biological productivity of the world's oceans comes from the less than 10 percent of their area that are fertilized with runoff from the coasts or continental shelf upwelling, with the vast open seas left a virtual desert. As a result, the ocean's $CO_2$ levels simply increase, with no useful and some potentially seriously harmful results.

What to do? The conventional answer from most of the global warming activist community has been to propose increased taxes on fuel and electricity, thereby dissuading people of limited means from making much use of such amenities. This program seems to me to be both unethical and impractical, and regardless of anyone's opinion on the matter, it is quite clear that it is failing to impact the growth of global carbon emissions in any significant way.

The basis for a much more promising approach was demonstrated by the British Columbia-based Haida native American tribe, who in 2012, launched an effort to restore the salmon fishery that has provided much of their livelihood for centuries. Acting collectively, the Haida voted to form the Haida Salmon Restoration Corporation, financed it with $2.5 million of their own savings, and used it to support the efforts of American scientist-entrepreneur Russ George to demonstrate the feasibility of open-sea mariculture through the distribution of 120 tons of iron sulfate into the northeast Pacific to stimulate a phytoplankton bloom which in turn would provide ample food for baby salmon.

By 2014, this controversial experiment proved to be a stunning, over the top, success. In that year, the number of salmon caught in the northeast pacific more than quadrupled, going from 50 million to 219 million. In the Fraser River, which only once before in history had a salmon run greater than 25 million (about 45 million in 2010), the number of salmon increased to 72 million.

"Up and down the West Coast fisheries scientists and fishers are reporting they are baffled at the miraculous return of salmon seen last fall and expected this year" commented George. "It is of course all because when we take care of our ocean pasture. Replenish the vital mineral micronutrients that we have denied them through our high and rising $CO_2$ just one old guy (me) with a dozen Indians can bring the ocean back to health and abundance."[4]

In addition to producing salmon, this extraordinary experiment yielded a huge amount of data. Within a few months after the ocean fertilizing operation, NASA satellite images taken from orbit showed a powerful growth of phytoplankton in the waters that received the Haida's iron.[5] It is now clear that as hoped, these did indeed serve as a food source for zooplankton, which in turn provided nourishment for multitudes of young salmon, thereby restoring the depleted fishery, and providing abundant food for larger fish and sea mammals as well. In addition, since those diatoms that were not eaten went to the bottom, a large amount of carbon dioxide was sequestered in their calcium carbonate shells.

Unfortunately, the experiment, which should have received universal acclaim, was denounced by many leading environmental activists. For example Silvia Ribeiro, of the international environmental watchdog ETC group, objected to it on the basis that it might undermine the case for carbon rationing. "It is now more urgent than ever that governments unequivocally ban such open-air geoengineering experiments. They are a dangerous distraction providing governments and industry with an excuse to avoid reducing fossil fuel emissions."

Writing in the New York Times, Naomi Klein, the author of a book on "how the climate crisis can spur economic and political transformation,"[6] said that "At first, it felt like a miracle." But then she was struck by a disturbing thought: "If Mr. George's account of the mission is to believed, his actions created an algae bloom in an area half of the size of Massachusetts that attracted a huge array of aquatic life, including whales that could be 'counted by the score.'...I began to wonder: could it be that the orcas I saw were on the way to the all you can eat seafood buffet that had descended on Mr. George's bloom? The possibility...provides a glimpse into the disturbing repercussions of geoengineering: once we start deliberately interfering with the earth's climate systems – whether by dimming the sun or fertilizing the seas – all natural events can begin to take on an unnatural tinge. ...a presence that felt like a miraculous gift suddenly feels sinister, as if all of nature were being manipulated behind the scenes."

But the salmon are back.

Not only that, but contrary to those who have denounced the experiment as reckless, its probable success was predicted in advance by leading fisheries scientists. "While I agree that the procedure was scientifically hasty and controversial, the purpose of enhancing salmon returns by increasing plankton production has considerable justification," Timothy Parsons, professor emeritus of fisheries science at the University of B.C. told the Vancouver Sun in 2012. According to Parsons, the waters of the Gulf of Alaska are so nutrient poor they are a "virtual desert dominated by jelly fish." But iron-rich volcanic dust stimulates growth of diatoms, a form of algae that he describes as "the clover of the sea." As a result, volcanic eruptions over the Gulf of Alaska in 1958 and 2008 "both resulted in enormous sockeye salmon returns."

The George/Haida experiment is of historic significance. Starting as a few bands of hunter-gatherers, humanity expanded the food resources afforded by the land a thousandfold through the development of agriculture. In recent decades, the bounty from the sea has also been increased through rapid expansion of aquaculture, which now supplies about half our fish. Without these advances, our modern global civilization of 8 billion people would not be possible.

But aquaculture makes use only of enclosed waters, and commercial fisheries remain limited to the coasts, upwelling areas, and other small portions of the ocean that have sufficient nutrients to be naturally productive. The vast majority of the ocean, and thus the Earth, remains a desert. The development of open sea mariculture could change this radically, creating vast new food resources for both humanity and wildlife. Furthermore, just as increased atmospheric carbon dioxide levels have accelerated the rate of plant growth on land, so increased levels of carbon dioxide in the ocean could lead to a massive expansion of flourishing sea life – more than fully restoring the world's depleted wild fisheries – provided humans make the missing critical trace elements needed for life available across the vast expanse of the oceans.

The point deserves emphasis. The advent of higher carbon dioxide levels in the atmosphere has been a boon for the terrestrial biosphere, accelerating the rate of growth of both wild and domestic plants, and thus expanding the food base supporting humans and land animals of every type. Yet in the ocean, increased levels of carbon dioxide not exploited by biology could lead to acidification. By making the currently barren oceans fertile, however, mariculture would transform this apparent problem into an extraordinary opportunity.

Such an effort would more than suffice to limit global warming. Indeed, were a few percent of the Earth's open ocean deserts fully enlivened by mariculture, the entirety of humanity's current $CO_2$ emissions would be turned into phytoplankton, with available worldwide fish stocks greatly increased as a result.[7]

The situation is ironic. In some places in the ocean, excessive nutrients delivered by runoff of agricultural fertilizers cause local algae blooms that are so massive as to destroy all other aquatic life. Yet, when delivered in the right amounts, such "pollutants" become the key to creating a vibrant marine ecology.

You can irrigate a farm to make it productive, or you can flood it and destroy the crops. You can fertilize land with horse manure, or you can...well, you get the idea.

"Pollution" is simply the accumulation of a substance which is not being put to good use. Carbon dioxide emissions are neither good nor bad in themselves. They are good for parts of the biosphere that are ready to make use of them, and bad for those that are not.

Say what you will, humans are not going to stop using fossil fuels any time soon. So we need to prepare the Earth to take full advantage of the resulting emissions.

That understood, what do we do?

The essence of the right strategy is not to try to deny access to fossil fuels. That course is neither moral nor practical. The right strategy is to put the emissions resulting from fossil fuel to good use. This can be done by making use of nuclear power to expand and quicken the biosphere.

The land-based biosphere is already exploiting the $CO_2$ enrichment of the atmosphere, but, with our help, it could do much more. Carbon dioxide is just one of the two fundamental resources supporting life on Earth. The other is water.

Most of the Earth's land does not receive sufficient rainfall to support anything remotely resembling its full potential for plant growth. This can be appreciated by simply comparing the biological productivity of different regions of the Earth, which differ by many orders of magnitude. If you fly from Denver to Los Angeles and take the trouble to look out the window you will see that half of the United States is actually an uninhabited desert. Similar sights can be witnessed while

**Fig. 14.3.** Photographs of the Earth taken from space. Most of the planet's land is desert. (Credits: NASA)

overflying most of the land areas of the planet. For example, see the photographs of the Earth taken from space shown in fig. 14.3. Note how not only the American west, but most of Africa, the Middle East, Central Asia, and Australia are all deserts. Furthermore, even the green parts are much less productive than they could be, as anyone can see who witnesses the difference between an irrigated piece of farmland and adjacent unirrigated land. Put simply, the Earth lacks sufficient fresh water.

Using nuclear power we can remedy this defect. The Earth has no shortage of salt water, and all it takes to turn it into fresh is energy. Nuclear power can readily meet this need. In fact, the waste heat from nuclear power plants can be used to desalinate massive amounts of sea water, with almost no cost to the stations' grid power at all. Today Australia is a desert continent, virtually uninhabited except for fertile fringes along its coasts. There is no reason why this must be so. Using mass produced nuclear power plants for desalination, not only Australia but even vaster tracts of desert stretching across Asia, the Middle East, Africa, the sub-Arctic, and the Americas could be turned into fertile land, filled with grand forests, delightful meadows, towns, and farms, producing food, inventions, and thoughts, and hosting abundant and diverse wildlife.

Then there are the oceans, which are largely desert as well, not because of a shortage of $CO_2$ or water, to be sure, but of trace nutrients, like iron, phosphorus, and nitrates that come from the land. In fact however, these are only lacking in the surface waters of the oceans. Below a few hundred meters in depth they are present in abundance, because the deeper waters, lacking sunlight, have not been scavenged for their nutrients by photosynthetic organisms. If these nutrient rich deep waters could be pumped to the surface, they could turn every part of the vast desert ocean into nurseries for life as productive as natural upwelling areas such as the Grand Banks. By employing floating or island based nuclear power plants for this purpose, the abundance of life in the world's oceans could be multiplied tenfold or more.

This glorious profusion of life both on land and sea would act to limit the rise of $CO_2$ concentration in the atmosphere, even in the face of expanded use of fossil fuels by an increasingly prosperous humanity.

Previous human societies existed under the condition that most of their members lived in slavery or poverty. Their leading citizens could only bear that shame by looking the other way. It has become a popular pastime in recent years to condemn them for this shortcoming, even as we today continue it. But this need not be so. Our prosperous civilization need not exist in parallel with, despite, or owing to, the impoverishment of others. We can create a world where no one is denied the chance for happiness, and where moreover we can stand proud in what we have done for the planet and its community of life which gave us birth. We now live on fertile islands in a planet of dry and salt water deserts. If we unleash our creativity, we can make those deserts bloom. We can prove here, and then on Mars and many worlds beyond, that we are not the enemies of life, but the vanguard of life.

We can continue the work of creation. In doing so, we shall make a profound statement as to the precious worth of the human race, and every member of it.

No one will be able to look upon our works and not be prouder to be human.

# CHAPTER 15

# THE WAY FORWARD

**NUCLEAR POWER CAN** provide the energy for an unlimited and magnificent human future. But the technological revolution it offers has thus far been strangled by political constraints, mismanagement, poor decisions, and outright sabotage. How can this situation be rectified?

There are four areas that need to be addressed. These include regulatory reform, waste disposal, support for research and development, and public understanding.

Let's talk about each of them.

## REGULATORY REFORM

The most important thing that needs to be done to provide humanity with the benefits of nuclear power is regulatory reform. When antinuclear activists claim that nuclear power is a failure because it simply costs too much they are lying. In fact, it is the activists themselves who have multiplied the costs of nuclear power by creating and exploiting a system of mendacious hyper regulation. They are like a poisoner on trial who claims that his victim died of heart failure.

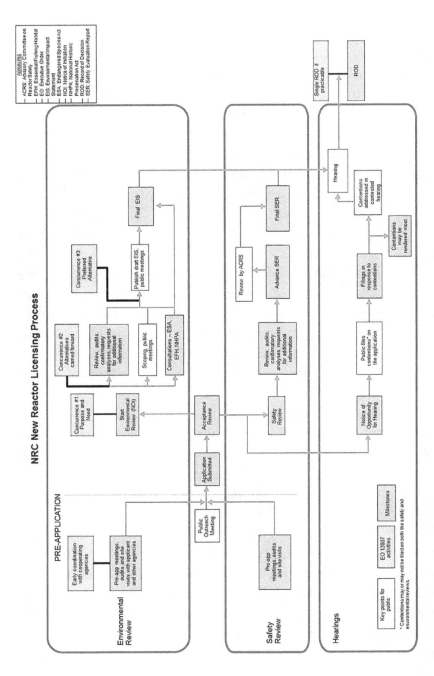

**Fig. 15.1** The NRC Nuclear power plant licensing process[1]. (Credit: US NRC)

Relative to the energy it provides, the fuel for nuclear reactors is extremely cheap, comprising only about 5 percent of total power costs. It is plant financing and construction costs that dominate the cost of nuclear electricity. These are largely determined by the time it takes to complete a plant construction project, which in turn is controlled by the regulatory process.

The insane nature of the process governing the building of a nuclear power plant is partially shown in figure 15.1, which depicts the Nuclear Regulatory Commission (NRC) 32-step construction licensing process.[1] I say partially shown, because many of these steps require input drawn from similar multi-step processes undertaken by other local, state, and federal agencies, notably the Environmental Protection Agency (EPA.) Each of these hundreds of steps not only requires the time taken by the frequently slow-moving agencies involved, but is open to legal intervention by "the public," that is, lawyers representing groups committed to stopping the plant. These employ the numerous opportunities provided to them to both halt and vastly increase the cost of the process by throwing it into the courts.

For example, in order to get its construction license, the utility must first perform an Environmental Assessment for the NRC. This can take a year or so. The NRC, then, basing itself on this data but requiring more, as well as the same data updated or in an alternative form, will then draw up an Environmental Impact Statement (EIS) for evaluation by the EPA. By law, the NRC must write the EIS within two years. In fact, however, the NRC operates without constraint by law, and actually takes an average of four years to write the EIS, and sometimes as long as six. The EPA, which is thoroughly infested with antinuclear activists itself, will then take its time evaluating the EIS, coming up with demands for more information. These will not only include matters nominally related to plant or public safety, but things entirely outside such a purview. For example, it is not uncommon for the EPA to demand a comprehensive study justifying the selection of nuclear power for the plant, comparing it to all possible alternatives, including gas, coal, oil, solar, wind, hydroelectric, cogeneration, or conservation.

Imagine the situation: You buy some land and decide to build a log cabin on it. But then when you go to the local authority to get a building permit, they ask you not just for your construction plans, but proof that a log cabin should be built on your land, rather than an A-frame, a chalet, a ranch house, a Cape Cod, a barn, an apartment building, a candy store, a gas station, a zoo, an anti-ballistic missile defense base, or nothing at all! Then, if somehow you provide such proof to the satisfaction of the local authority, your lifelong enemy goes to court and challenges the validity of that approval, forcing you to hire a lawyer, endure three years of discovery motions, and then throw the dice to try to win the case in court. If you win, the other side will appeal, and you will need to go to trial again.

That is what the regulatory process governing the construction of a nuclear power plant is like. This process was created by the antinuclear Carter administration in the 1970s. In the early 1970s it took an average of four years to build a nuclear power plant in the United States. With the advance of technology and the gaining of experience, that time should have been cut to two years by now. Instead, as a result of the implementation of this bizarre process, it now takes sixteen.

Experience has shown that the cost of building a nuclear power plant increases roughly in proportion to the construction time squared. This is because the longer the project goes on, the more requirements, technical changes, and legal actions are levied on it. The slower the project moves, the more hits it takes. By multiplying the time it takes to complete a nuclear power plant, the antinuclear regulatory process has inflated the cost of nuclear power by two orders of magnitude.

But that's just the beginning. After construction is completed, the plant must again go through a similar process to obtain an operating license. If this cannot be obtained, or is put in abeyance by capricious politicians allied to the antinuclear movement (as New York governor Mario Cuomo did to the $5 billion Shoreham nuclear plant in the 1980s),[2] the entire project becomes a complete financial loss. This element of risk has greatly increased the cost of financing nuclear power plants.

The nuclear industry has not been the only victim of this process. By stopping the building of nuclear power plants or radically increasing their costs, the antinuclear activists have served the cause of protecting other more expensive (and much more polluting) power sources. This is why such groups have been funded to conduct their wrecking campaigns against nuclear power, with the generous donors ranging from oil and coal companies in the 1970s to "renewable energy" bandits in the more recent period.[3] The resulting increased costs of electric power have been born by the public, not only directly through increased utility bills, but more importantly though added costs borne by industry, inflating the costs of all its products, decreasing competitiveness, and cutting into wages and job.

There has also been a massive environmental and public health cost as well. If the US nuclear industry had been allowed to continue to grow at the same rate it was going in the 1960s and early 1970s, the American electric grid would long since have eliminated its use of fossil fuels (as did that of France, 75% of whose electricity comes from nuclear power.) Instead, the imposition of mendacious hyper regulation stopped the growth of the industry dead in its tracks.

In short, radical regulatory reform is required.

The current absurd system should be replaced with one in which the utility applies to the NRC for a combined construction and operating license. The NRC would then have two years to grant the license, demand corrections, or present grounds for refusal. With the license granted and the plant built, operations should then be allowed to proceed following a final on-site plant safety inspection by the NRC. There should be no Environmental Impact Statement, and no requirement to explain why the utility did not decide to do something else. The EPA should not be involved in the permitting process at all. Rather its role should be limited to holding utilities to account should they actually cause any emissions to the environment. The "public," i.e. lawyers for outsiders who wish to destroy the project, should have no role in the process whatsoever.

Limiting the EPA role to environmental enforcement rather than permitting may seem wildly radical to some, but it actually common sense. When you plan a road trip, you are not required to go to the police in advance and prove to their satisfaction that you will not speed. Rather, you just go, and if you speed, the police give you a ticket. That is how regulations should be enforced. Indeed, that is how all laws should be enforced. In any civilized society, people should be assumed innocent until proven guilty, and certainly no one should be arrested for a crime that hasn't happened yet.

Eliminating the power of hostile outsiders to intervene in the permitting process is not merely a practical necessity to enable nuclear power, it is necessary to maintain the basic principles of liberty under law. As explained in America's Declaration of Independence, human beings are endowed with inalienable rights, "including life, liberty, and the pursuit of happiness, and *to ensure these rights governments are instituted among men.*" This is a very important point. The legitimate and essential purpose of government is to protect individual rights, including the right of individuals to enjoy the use of their own property. Contrary to some libertarians, there is no such thing as private property without a state. You don't own your house because you live in it. No matter how personally formidable you may be, a gang of thugs of sufficient size could readily eject you from your home. No, you own your house because the armed forces of the state, starting with the local police but backed up, if necessary, by the National Guard and the United States Army, are ready and available to defend your property title. That ultimately is why we have a government. It protects each of us from anarchic oppression by everyone else. Without it, none of us would own anything, and none of us would be safe.

However, placing this awesome power at the disposal of private enemies is directly contrary to this principle. Your home is your castle. The police exist to protect your right to enjoy it, not to fulfill the desires of your enemies to remove it. If someone thinks Americans should be using solar energy in place of the nuclear power you offer, it is legitimate for them to build a solar energy plant to compete with you. It is not legitimate for them to attempt to make use of the government's regulatory power to shut you down. Anyone found to be connected to such outside interested parties needs to be expelled from the NRC, and certainly no one openly affiliated with them should be allowed to participate in the regulatory process at any step.

If you drive around Moscow, you will see many half-completed hotels whose construction was stopped when their competitors, exercising influence with the government, managed to get their building permits pulled. The same corrupt way of doing business has become standard practice in the nuclear power plant permitting process. It needs to end.

A final issue associated with regulation is financing. Right now Russia and China are cleaning the clock of American nuclear industry by providing aggressive financing of nuclear power plant orders around the world. The United States has an institution known as the Export-Import Bank whose mission is to provide precisely such competitive financing for American companies seeking to sell technology abroad. But it is not providing this service for the nuclear industry. That needs to change. The World Bank has a regulation preventing it from providing financing nuclear plants in the developing sector. That ban needs to be lifted as well.[4]

## WASTE DISPOSAL

When the Sierra Club in 1974 announced its opposition to nuclear power, it identified preventing the safe disposal of nuclear waste as a key tactic to use to wreck the nuclear industry.[5] As a result of this malicious campaign, nuclear waste reprocessing, subsea bed disposal, and land based disposal have all been blocked, forcing utilities to store their radioactive waste on site. This has added costs to the utilities operations which have been passed on to the public both through higher rates and through higher taxes to compensate utilities for these costs, as required by law. Furthermore, storing nuclear waste on sites near major metropolitan areas could, under worst-case scenarios – such as Fukushima – expose the public to dangers of radiological release that would be quite impossible if the waste was stored in remote areas. There is no technical obstacle to either nuclear waste processing (the French do it) or land-based disposal (the US military has been storing its waste since 1999 in salt formations in the Waste Isolation Pilot Plant near Carlsbad, New Mexico.)[6] Sub seabed disposal, which involves glassifying the waste so it is not water-soluble, putting it in stainless steel canisters, and dropping into mid-ocean sediments that have been stable for hundreds of millions of years, is even more straightforward. Once buried under hundreds of meters of mud in the middle of the ocean, the waste isn't going anywhere, and no nomads roaming the world after the next ice age obliterates our civilization will ever be endangered by it, regardless of how poorly informed they may be.

So when antinuclear spokesmen say there is no safe way to dispose of nuclear waste, they are lying. The truth is that they have prevented safe disposal of nuclear waste. In doing this, they have partially achieved their purpose, as about a dozen states have passed laws banning construction of new nuclear power plants "until a safe way to dispose of the waste can be found."

After abandoning nuclear waste reprocessing and sub sea-bed disposal, the United States Department of Energy focused its radioactive waste disposal work on the option of storing the waste in a repository under Yucca Mountain in Nevada. I'm not a big fan of this choice, because I think that waste reprocessing followed by sub seabed disposal would be a better way to go. Nevertheless, it clearly would be wiser and safer to store nuclear waste under a mountain in the desert than in the suburbs of major cities. Despite all their posturing about putative concerns as to what might happen in the desert in future ages, the organizations campaigning to block the Nevada option are clearly not concerned about public safety. Rather they are attempting to maximize the danger, or perceived danger, posed by radioactive waste to the public in order prevent nuclear power.

The Biden administration says its believes that climate change is an "existential crisis." That means a crisis that *threatens human existence*. Yet, while saying it is interested in the potential contribution of nuclear power to solve this crisis, it has endorsed the environmentalist campaign to stop the Yucca Mountain project. Biden's DOE Secretary Jennifer Granholm says the DOE has no choice in this matter, because there is too much opposition to the program. Yet the opposition is being organized and led by Biden's own party.

Nuclear power cannot expand unless provision is made for the safe disposal of nuclear waste. The Democrats say *that human existence is at stake* in replacing fossil fuels with carbon-free power. That being the case, they should stop preventing this from being accomplished.

## SUPPORT FOR RESEARCH AND DEVELOPMENT

In recent years it has become fashionable in some circles to say that nuclear power would be a good thing, but only if we had more advanced systems than the pressurized water reactors that have been the industry mainstay since Rickover introduced them in the 1950s.

I do not agree with this point of view. The PWRs have been a resounding success. In the face of operation of close to one thousand units on land and sea over more than six decades, not a single person in the entire world has ever been seriously harmed, let alone killed, by a radiological release from a PWR. No other major industrial or energy technology can come close to matching that safety record.

Nevertheless, it would be a very good thing if nuclear technology were advanced further. Breeder reactors could multiply our nuclear fuel resources a hundredfold. Small modular reactors could open up new markets inaccessible to large PWRs, and potentially make reactors much cheaper by enabling mass production in factories. High temperature gas-cooled reactors and molten salt thorium reactors both hold great promise. New types of fission reactors for space applications are needed. The promise of thermonuclear fusion needs to be explored and developed.

These potentials need to be brought to fruition, and it is right and proper that government should help do so. In the case of advanced fission reactors, the Biden administration, to its credit, has continued the previous administration's initiative to foster entrepreneurial development of advanced nuclear concepts by issuing a fair number of research and development contracts in the tens of million dollars range.

So far, so good. But what the new entrepreneurial companies need most is not cash but facilities to test their concepts and regulatory reform to allow the reactors to be licensed and sold. If these are provided, investment money will follow.

The required facilities are available at places like Oak Ridge, Los Alamos, Livermore, Idaho National Lab, Hanford, the Nevada test site, and other federal government nuclear reservations. These should be made available on a no cost basis to entrepreneurial fission companies. Furthermore, there is considerable expertise in nuclear technology, neutronics, and related fields at the DOE's national labs. These should be made available for hire by the entrepreneurial nuclear companies on reasonable commercial terms.

Finally there is fusion. The United States severely damaged its fusion program by eliminating serious support of all non-tokamak concepts in the 1980s, and then killed it completely by aborting development of any new major American tokamaks in the 1990s. Instead the whole program was collapsed into support of the glacial ITER effort.

While the United States should participate in ITER, it needs a vibrant national fusion program as well. The entire US magnetic fusion budget for FY 2022 was $675 million, which is about 3% of the NASA budget, or 0.01% of the federal budget overall. At a minimum, the magnetic fusion budget should be tripled. This would allow the USA to build a tokamak (probably a spherical tokamak with a high magnetic field) capable of reaching ignition, as well as provide healthy support for diverse efforts exploring promising alternative concepts including field reversed configurations, spheromaks, and other advanced systems that take advantage of the collective self-organizing properties of plasmas. In addition, such a proper level of funding would allow the DOE to substantially support the growing entrepreneurial fusion efforts with matching funds, research grants, and in-kind assistance from the national labs.

## PUBLIC UNDERSTANDING

Finally, what is needed most is public understanding. This is an area where you, dear reader, can take an active hand.

The putative environmentalists have been able to institute a regulatory blockade against nuclear energy though a malicious and sustained campaign of disinformation, distortions, and fear. Through their lies, they have denied humanity enormous benefits, and if allowed to continue, will put chains on the future. If they have their way, the hopes of billions of people to escape from brutal poverty will be denied in the name of a false imperative to constrain human aspirations to "save the planet."

These patrons of ignorance need to be countered. Indeed, they need to be exposed for the frauds they are. As someone who now knows the truth of these matters, you need to speak up.

This will not make you popular. Support for nuclear power is politically incorrect. In many circles, it will make your socially unacceptable. But this is a battle of ideas that needs to be fought, and won.

A few years before he died, I had the opportunity to ask Rickover what he thought the most important personal character an engineer needed to have. He answered "courage."

By that he did not mean physical courage, for example the willingness to charge a machine gun nest. He meant intellectual courage, the willingness to hold one's ground and speak truth, regardless of whether that allows you to fit in with the team.

I think he was right. Certainly that aspect of his character was what allowed him to accomplish so much. In my own life, I have tried to hold to that advice, and to the extent I have been able to make some contributions, I believe it has been due to that commitment.

It's not easy. Expulsion from the congregation is the price you must pay for the right to speak freely. It takes a brave soul to adopt such a course.

But nothing great has ever been accomplished without courage.

## FOCUS SECTION: THE FISSION FREEDOM FIGHTERS

You can fight for fission on your own. But if you prefer, you can also do it in the company of others. Should organized action be to your taste, there are a host of outfits of many political stripes that you can sign up with.

FDR's Democratic Party created nuclear power, and for the quarter century following World War II was its main source of political support. However, since the Democrats reversed themselves in the 1970s to embrace reactionary Malthusian antihumanism, it has been the Republican Party and its band of think tanks and conservative organizations that has been the nuclear industry's primary political ally.

On the industry side these forces have included the American Nuclear Society, the Nuclear Energy Institute, the Nuclear Innovative Alliance, and, with respect to fusion, Fusion Power Associates and the Fusion Energy Association. Noteworthy conservative supporting organizations include the Heritage Foundation, the American Action Forum, the center-right Clear Path, and the small but fiercely combative Doctors for Disaster Preparedness.

These, and similar groups, in alliance with the utilities and industrial partners themselves have been the principal defenders of nuclear power in the United States for the past four decades. Their efforts, however, have proved insufficient, serving at best as a rearguard defense for an industry in retreat.

It has therefore been greatly encouraging to see in recent years the reemergence of an increasingly vocal pro nuclear faction within the Center Left. Prominent leaders of this movement have included the Breakthrough Institute[7] and a Democratic Party thinktank called the Third Way.[8]

Founded in 2003 by Ted Nordhaus and Michael Shellenberger, the Breakthrough Institute has gathered an impressive array of Enlightenment humanist intellectuals including sociologist Bruno Latour, journalist and author Gwyneth Cravens, Nobel Prize-winning physicist Burton Richter, political and environmental scientist Roger A. Pielke Jr., sociologist Dalton Conley, Oxford professor Steve Rayner, plant geneticist Pamela Ronald, sociologist Steve Fuller, environmental thought leader Stewart Brand, philosopher Steven Pinker, and ecologist Emma Marais.

In 2015 this group issued an "Ecomodernist Manifesto" calling for humanistic, non-zero-sum approaches to solving environmental problems, including climate change. Taking a bold stand in defiance of established left wing Malthusian groupthink, the Ecomodernist Manifesto called for nuclear power.[9]

"Human civilization can flourish for centuries and millennia on energy delivered from a closed uranium or thorium fuel cycle, or from hydrogen-deuterium fusion," it proclaimed. The Manifesto then went on to say:

"Nuclear fission today represents the only present-day zero-carbon technology with the demonstrated ability to meet most, if not all, of the energy demands of a modern economy. However, a variety of social, economic, and institutional challenges make deployment of present-day nuclear technologies at scales necessary to achieve significant climate mitigation unlikely. A new generation of nuclear technologies that are safer and cheaper will likely be necessary for nuclear energy to meet its full potential as a critical climate mitigation technology. In the long run, next-generation solar, advanced nuclear fission, and nuclear fusion represent the most plausible pathways toward the joint goals of climate stabilization and radical decoupling of humans from nature....The ethical and pragmatic path toward a just and sustainable global energy economy requires that human beings transition as rapidly as possible to energy sources that are cheap, clean, dense, and abundant."

**Fig. 15.2** Left leaning fission freedom fighters. (Left) The American Good Energy Collective's Jessica Lovering. (Right) The UK's Kirsty Gogan, founder of Energy for Humanity.

Breakthrough Institute staffer astrophysicist Jessica Lovering followed this up with a series of policy papers identifying specific areas for action to break the nuclear deadlock.[10,11,12] Lovering then went on become a fellow at Energy for Growth[13] and founded the Good Energy Collective,[14] which published a multitude of additional policy papers calling for advanced nuclear power and making the case for the necessity of using nuclear energy to lift the world's developing nations out of poverty.

Many of the leaders, experts, and spokespersons of the nuclear lefties have been women. This wasn't a completely new development, as Marie Curie and Lise Meitner had founded nuclear physics by discovering radioactivity and nuclear fission respectively. Yet it must be said that that the arrival of a force of fierce female fission freedom fighters on the political battlefield has had a real impact, reshaping the nuclear message into a form congenial to progressives.

Their recommendations have begun to make their way into Democratic Party policy circles, with a key role being the think-tank known as the Third Way.

In 2016, the smart money hit the canvas when Donald Trump defeated Hilary Clinton in the November presidential election. Clinton had a great resume and the full and enthusiastic backing of the nation's political establishment and nearly all news and entertainment media. By any conventional calculation, she should have beat the erratic Trump by twenty points. Instead, by the time election night was over, Trump had won. While many of Clinton's disappointed supporters sought solace in blaming the defeat on racism, sexism, or Putin, a cold hard look at the electoral map told a different story. Clinton lost the election in the industrial Midwest. Across the rest of the nation, the states were all won or lost in accord with the pollster's confident predictions. But to the amazement of all, Pennsylvania, Ohio, Michigan, and Wisconsin – the entire industrial heartland whose unionized labor had been the base of the Democratic Party since FDR – was swept by Trump.

The party clearly faced a problem, and at least some of its leaders recognized that the core of the issue was reconciling the passions of its environmentalist supporters with the real needs of blue-collar workers. "For what profitteth a candidate if she gains the donations of Tom Steyer but loses the votes of the industrial Midwest," commented one wag. There had to be a way to please both.

The Party was not about to abandon its core belief that carbon emissions represented an existential threat to humanity. So changing its position on coal mining or fracking was out. But nuclear power is carbon free. If the party embraced nuclear power, it could support both economic growth *and* environmental necessity. Not all Democrats saw things that way, but some did. Thus was born the Third Way.

Similar factions have begun to appear among center-left social democrats in Europe as well.[15]

Personally, I do not agree with the Third Way line that nuclear power is needed to stop the existential crisis of climate change. I don't believe there is such a crisis and I'm not willing

**Fig. 15.3** Mothers for a Nuclear Germany demonstrate in Berlin. "Geht mit Kernkraft" means "Go with Nuclear Power." (Mothers for Nuclear Deutschland Schweiz Österreich/<u>via Twitter</u>)

TO pretend I do to sell books. In the 1950s and sixties, Alvin Weinberg attempted to use the "existential crisis" of that time, the "Population Explosion," to make the case for nuclear power.[16] I think that was a mistake, because the Malthusian ideologues pushing the Population Crisis,[17] were intrinsically hostile to nuclear power. They hated it, for the same reason that the current Green anti human movement hates nuclear power: it threatens to solve a problem they need to have.

But peace. In policy, as in religion, there is more than one path to virtue. If their concern over global warming (or lost elections) convinces the center left to switch sides and fight for nuclear power, I'm cool with that. We can march separately but strike together.

But here's the thing. History is not a spectator sport. Things happen because people make them happen. If you want a positive human future, you need to fight for it.

Whether you are a leftist, a rightist, a centrist, or an individualist, there's a place in this fight for you.

# CHAPTER 16

# THEIR PROGRAM AND OURS

*The law of existence prescribes uninterrupted killing,*
*so that the better may live.*
— Adolf Hitler, 1941

**HUMAN CIVILIZATION CURRENTLY** faces many serious dangers. The most immediate catastrophic threat, however, does not come from environmental degradation, resource depletion, or even asteroidal impact. It comes from *bad ideas.*

Ideas have consequences. Bad ideas can have really bad consequences.

The worst idea that has ever been is that the total amount of potential resources is fixed. It is a catastrophic idea, because it sets all against all.

Currently, such limited-resource views are quite fashionable not only among futurists, but much of the body politic. But if they prevail, then human freedoms must be curtailed. Furthermore, world war and genocide would be inevitable, for if the belief persists that there is only so much to go around, then the haves and the want-to-haves are going to have to duke it out, the only question being when.

This is not an academic question. The twentieth century was one of unprecedented plenty. Yet it saw tens of millions of people slaughtered in the name of struggle for existence that was entirely fictitious. The results of similar thinking in the twenty-first could be far worse.

The logic of the limited resource concept leads down an ever more infernal path to the worst evils imaginable. Basically it goes as follows:

1.   Resources are limited. Therefore:
2.   Human aspirations must be crushed.
3.   So, some authority must be empowered to do the crushing.
4.   Since some people must be crushed, we should join with that authority to make sure that it is those we despise who are the ones crushed, rather than us.
5.   By getting rid of such inferior people, we can preserve scarce resources and advance human social evolution, thereby helping to make the world a better place.

The fact that this case for oppression, tyranny, war and genocide is entirely false has made it no less devastating. Indeed, it has been responsible for most of the worst human-caused disasters of the past 200 years. So let's take it apart.

Two hundred years ago, the English economist Thomas Malthus set forth the proposition that population growth must always outrun production as a fundamental law of nature. This theory provided the basis for the cruel British response to famines in Ireland and India during the mid through late 19th Century, denying food aid or even taxation, rent, or regulatory relief to millions of starving people on the pseudoscientific grounds that their doom was inevitable.[1]

Yet, as the data in Figure 3.1 shows, the Malthusian theory is entirely counterfactual. In fact, over the two centuries since Malthus wrote, world population has risen sevenfold while

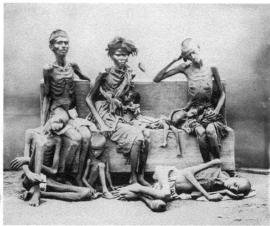

**Figure 16.1** Famine Victims in Ireland, 1846, and India, 1876. In response to pleas that the Empire relent on taxes and rents, Malthusian Governors in both cases claimed this would only make things worse by allowing their subjects to multiply.

inflation-adjusted global gross domestic product per capita has increased by a factor of fifty, and absolute total GDP by a factor of 350.

Indeed, it is clear that the Malthusian argument is fundamentally nonsense, because resources are a function of technology, and the more people there are, and the higher their living standard, the more inventors, and thus inventions, there will be—and the faster the resource base will expand.

Our resources are growing, not shrinking, because resources are defined by human creativity. In fact, there ar no such things as "natural resources." There are only natural raw materials. It is human ingenuity that turns natural raw materials into resources.

Land was not a resource until people invented agriculture, and it is human ingenuity, manifested in continuous improvements in agricultural technology, which has multiplied the size of that resource many times over.

Petroleum was not originally a resource. It was always here, but it was nothing useful. It was just some stinky black stuff that sometimes oozed out of the ground and ruined good cropland or pasture. We turned it into a resource by inventing oil drilling and refining, and by showing how oil could be used to replace whale oil for indoor lighting, and then, later, liberating humanity with unprecedented personal mobility.

This is the history of the human race. If you go into any real Old West antique store and look at the things owned by the pioneers you will see things made of lumber, paper, leather, wool, cotton, linen, glass, carbon steel, maybe a little bit of copper and brass. With the arguable exception of lumber, all of those materials are artificial. They did not, and do not, exist in nature. The civilization of that time created them. But now go into a modern discount store, like Target. You will see some items made of the same materials, but much more made of plastic, synthetic fibers, stainless steel, fiberglass, aluminum, and silicon. And in the parking lot, of course, gasoline. Most of the materials that make up the physical products of our civilization today were unknown 150 years ago. Aluminum and silicon are the two most common elements in the Earth's crust. But the pioneers never saw them. To the people of that time, they were just dirt. It is human invention that turned them from dirt into vital resources.

Twenty years ago shale was not a resource. Today, as a result of the invention of new techniques of horizontal drilling and fracking, it's an enormous resource. In the past ten years, we've used it to increase our oil production by 50 percent. In the past twenty years, America's gas reserves have tripled, and we can and will do that and more for the world at large.

But by far the greatest potential resources of all derive from tapping the binding energy of atomic nuclei. Uranium and thorium were not resources at all until we invented nuclear power, but their enormous promise is still being stifled. The energy that could be made available from the fusion of deuterium in sea water, let alone that of hydrogen with boron, is so vast as to be beyond all reckoning.

So the *fact* of the matter is that humanity is not running out of resources. We are exponentially expanding our resources. We can do this because the true source of all resources is not the earth, the ocean, or the sky. It is human creativity. It is people who are resourceful.

It is for this reason that, contrary to Malthus and all of his followers, the global standard of living has continuously gone up as the world's population has increased, not down. The more people, – especially free and educated people – the more inventors, and inventions are cumulative.

Furthermore, the idea that nations are in a struggle for existence is completely wrong. Darwinian natural selection is a useful theory for understanding the evolution of organisms in nature, but it is completely false as an explanation of human social development. This is so because, unlike animals or plants, humans can inherit acquired characteristics – for example new technologies – and do so not only from parents but from those to which they are entirely unrelated. Thus inventions made anywhere ultimately benefit people everywhere. Human progress does not occur by the mechanism of militarily superior nations eliminating inferior nations. Rather inventions made in one nation are transferred all over the world, where newly combined with other technologies and different mind sets, they blossom in radical new ways. Paper and printing were invented in China, but they needed to be combined with the Phoenician-derived Latin alphabet, German metal-casting technology, and European outlooks concerning freedom of conscience, speech, and inquiry to create a global culture of mass literacy. The same pattern of multiple sourcing of inventions holds true for virtually every important human technology today, from domesticated plants and animals to telescopes, rockets, nuclear power, and interplanetary travel.

Based on its inventiveness and its ability to bring together people and ideas from everywhere, America has become extremely rich, inciting envy elsewhere. But other countries would not be richer if America did not exist, or was less wealthy or less free. On the contrary, they would be immeasurably poorer.

Similarly America would not benefit by keeping the rest of the world underdeveloped. We can take pride in our creativity, but in fact we would be much better off if more people from other countries could have a chance to live, and thus contribute to progress, as we do.

Nevertheless, so long as humanity is constrained to a resource base that appears to be fundamentally limited – such as fossil fuels, whose known reserves divided by consumption is always reckoned in decades – the arguments of the Malthusians have the appearance of self-evident truth, and their triumph can only have the most catastrophic results.

Indeed, one has only to look at the history of the 20th Century, and the Malthusian/national social Darwinist rationale that provided the drive to war of both Imperial and especially Nazi Germany to see the horrendous consequences resulting from the widespread acceptance of such myths.

As the German General Staff's leading intellectual General Friedrich von Bernhardi put in in his 1912 bestseller *Germany and the Next War*; "Strong, healthy, and flourishing nations increase in numbers. From a given moment they require a continual expansion of their frontiers, they require new territory for the accommodation of their surplus population. Since almost every part of the globe is inhabited, new territory must, as a rule, be obtained at the cost of its possessors—that is to say, by conquest, which thus becomes a law of necessity."[2] Having accepted that war was inevitable, the only issue for the Kaiser's generals was when to start it, and they chose sooner rather than later so as not to give Russian industry a chance to develop.

Thus in 1914, the unprecedently prosperous European civilization was thrown into a completely unnecessary and nearly suicidal general war. A quarter century later, the same logic led the Nazis to do it again, with not merely conquest, but systematic genocide as their insane goal.

To be perfectly clear on this point, the crimes of the Nazis were not just committed in secret by a few satanic leaders while the rest of the good citizens proceeded with their decent daily lives in well-meaning ignorance. In point of fact, such blissful ignorance was not possible. At its height, there were over 20,000 killing centers in the Third Reich, and most were discovered by Allied forces within hours of their entry into the vicinity – as the stench of their crematoria made them readily detectable. Something on the order of a million Germans were employed operating these facilities, and several million more were members of armed forces or police units engaged in or supporting genocidal operations.[3] Thus nearly every German had friends or family members who were eyewitnesses to or direct perpetrators of genocide, who could, and did, inform their acquaintances as to what was happening. (Many sent photos home to their parents, wives, or girlfriends, depicting themselves preparing to kill, killing, or posing astride the corpses of their victims.) Moreover, the Nazi leadership was in no way secretive about its intent; genocide directed against Jews and Slavs was the openly stated goal of the Party *that 18 million Germans voted for* in 1932. On March 20, 1933, less than 2 months after the Nazi assumption of power, SS leader Heinrich Himmler made it clear that these voters would have their wishes gratified, by announcing the establishment of the first formal concentration camp, Dachau, *at a press conference.* Furthermore, the implementation of the initial stages of the genocide occurred in public, with systematic degradation, beatings, lynchings, and mass murder of Jews done openly for all to see in the Reich's streets from 1933 onward, with the most extensive killings, such as those of the November 9-10, 1938 Kristallnacht pogrom, celebrated afterwards at enormous public rallies and parties.

So the contention that the Nazi-organized holocaust took place behind the backs of an unwilling German population is patently false. Rather, the genocidal Nazi program was carried out – and could only have been carried out – with the full knowledge and substantial general support of the German public. The question that has bedeviled the conscience of humanity ever since then is; how could this have happened? How could the majority of citizens of an apparently civilized nation choose to behave in such a way? Some have offered German anti-Semitism as the answer. But this explanation fails in view of the fact that anti-Semitism had existed in Germany, and in many other countries such as France, Poland, and Russia to a sometimes much greater extent, for centuries prior to the holocaust, with no remotely comparable outcome.

Furthermore, the Nazi genocidal program was not directed just against Jews, but also at many other categories of despised people, including invalids, Gypsies, and the entire Slavic race. Indeed, the Nazis had drawn up a plan, known as the Hunger Plan, for depopulating eastern Europe, the Balkans, and the Soviet Union through mass starvation following their anticipated victory, on the insane supposition that by ridding the land of its farmers they could make more food available.[4] It should be noted that the partial implementation of this plan in occupied areas not only caused tens of millions of deaths, but contributed materially to the defeat of the Third Reich, as it made it impossible for the Nazis to mobilize the human potential of the conquered lands on their own behalf. But not even such clear practical military and economic considerations could prevail against the power of a fixed idea.

In other words, as the Nazi leadership itself repeatedly emphasized, the genocide program was not motivated by mere old-fashioned bigotry. It certainly took advantage of such sentiments among rustics, hoodlums, and others to facilitate its operations. But it required something else to convince a nation largely composed of serious, solid, dutiful, highly literate, and fairly intellectual people to devote themselves to such a cause. It took Malthusian pseudo-science.[5]

Hitler himself was perfectly aware of the central importance of such an ideological foundation for his program of genocide. As noted holocaust historian Timothy Snyder wrote in in a September 2015 New York Times op ed; "The pursuit of peace and plenty through science, he claimed in 'Mein Kampf,' was a Jewish plot to distract Germans from the necessity of war."[6,7]

Once again, to be clear, the issue is not whether fusion power plants will be made available to humanity in large numbers in the proximate future. Rather it is how we, in the present, conceive the nature of our situation in the future. In reality, Nazi Germany had no need for expanded living space. Germany today is a much smaller country than the Third Reich, with a significantly higher population, yet Germans today live much better than they did when Hitler took power. So in fact, the Nazi attempt to depopulate Eastern Europe was completely insane, from not only a moral, but also a practical standpoint. Yet driven on by their false zero-sum beliefs, they tried anyway.

If it is allowed to prevail in the twenty-first century, zero-sum ideology will have even more horrific consequences. For example, there are those who point to the fact that Americans are four percent of the world's population, yet use 25 percent of the world's oil. If you were a member of the Chinese leadership and you believed in the limited resource view (as many do—witness their brutal one-child policy), what does this imply you should attempt to do to the United States?

On the other hand, there are those in the US national security establishment who cry with alarm at the rising economy and concomitant growing resource consumption of China.

**Fig. 16.2** Self-fulfilling prophesies. In 1912 the theoreticians of the German General Staff said it was inevitable that Germany would have to wage war for living space. In 2001, American geostrategists proclaimed we would need to fight for oil. Both were wrong. Both set the stage for disaster. Much worse could be on the way if such zero-sum ideology is not discredited.

They project of future of "resource wars," requiring American military deployments to secure essential raw materials, notably oil.[8]

As a result of acceptance of such ideology, the United States has initiated or otherwise embroiled itself in conflicts in the Middle East costing tens of thousands of lives (to us – hundreds of thousands to them) and trillions of dollars. For one percent of the cost of the Iraq war we could have converted every car in the USA to flex fuel,[9] able to run equally well on gasoline or methanol made from our copious natural gas. For another one percent we could have developed fusion power. Instead we are fighting wars to try to control oil supplies that will always be sold to the highest bidder no matter who owns them.

There were no valid reasons for the first two World Wars, and there is no valid reason for a third. But there could well be one if zero-sum ideology prevails. Despite the bounty that human creativity is producing, there are those in America's national security establishment who today are planning for resource wars against peoples who could and should be our partners in abolishing scarcity. Their equivalents abroad are similarly sharpening their knives against us. This ideology threatens catastrophe.

Today there is a dangerous new anti-Western, anti-freedom movement centered in Russia led by fascist philosopher Aleksandr Dugin, who is attempting to expand it worldwide. (He is doing so with significant success. The American "Alt Right" and a host of similar European "identarian" nativist movements all draw heavily from Dugin's ideas.[10] The basic concept is to Balkanize the West, strangle its economy, and undermine its commitment to Enlightenment humanist ideals by invoking tribalism, Malthusianism, and other reactionary notions drawn from a combination of brown, green, and red sources.) It is the contention of the Duginites that the world would be better off without America, or any other country with liberal values. Indeed, I was present at a conference on global issues held at Moscow State University, Dugin's home turf, in October 2013, when one of his acolytes got up and gave a fiery speech denouncing America for its profligate consumption of the world's resources, including its *oxygen* supply.[11] Such ideas amount to a call for war.

In order to triumph in that "clash of civilizations," Dugin calls for reestablishing the Russian Empire, with all its previous possessions including Ukraine, the Baltic States, Poland, and the states of the Caucasus, and central Asia, as the ruling core of a totalitarian "Eurasian Union" stretching "from Lisbon to Vladivostok." This superpower would then be able to defeat the world's remaining liberal democracies, acquiring control of all the world's supposedly finite resources and putting an end to humanity's dangerous experiment with progress and freedom.

The mixture of "left wing" and "right wing" antiliberal elements within Dugin's ideology have led some to perceive it as something fundamentally new. On the contrary, it is essentially a close remake of Nazism. Hitler's key political insight was that there is no contradiction between nationalism and socialism. On the contrary, invoking the tribal instinct is essential to arouse the passion necessary to implement the full collectivist agenda in its most lethal form. Thus he accurately named his movement "national socialism," abbreviated Nazism. Duginism builds on this, but aiming for a wider marker, shuns Nazism's obsession with the Nordic race per se. Instead it allows any race to conceive of itself as the master race, entitled to crush all others as well as deviant or disloyal individuals within its own ranks. Embracing the ideas of Nazi legal theorist Carl Schmitt, it denounces the idea of individual human rights as an Anglo-Saxon conceit that denies the freedom of the collective to work its will without restraint. Like Nazism, it also draws extensive elements from communism, environmentalism, and weird forms of primitive mysticism to synthesize an intellect-destroying cult that people might be willing to die for.[10]

In the 1920s Hitler was viewed by ruling circles in Germany as a kook, but a potentially useful one. Thus, during his imprisonment following his failed 1923 Beer Hall Putsch, Hitler was visited by Professor General Karl Haushofer, who instructed the young demagogue in the ideas of geopolitics, positing the need to unite the conservative Eurasian continental "Heartland" against the liberal cosmopolitan maritime "Rimland" to win the coming war of civilizations that would determine the human future. Duginism embraces this geostrategic ideology in full, with the only difference being that it moves the capital of the future Heartland world empire from Berlin to Moscow.

After Hitler took power, some members of the foreign policy establishments in other countries took the trouble to read his turgid opus *Mein Kampf*, but they generally dismissed it as irrelevant nonsense since it was clearly insanity.

What they missed was that it was a form of insanity useful to those who believed in the necessity for war and tyranny. Those who dismiss Duginism as merely the ravings of a crank are making the same mistake.

Dictators don't take orders from professors, but they do get their ideas from them. They also make use of the ideological constructs created by intellectuals to gain broader acceptance of their own agendas. While revanchist Russian dictator Vladimir Putin certainly doesn't take orders from Aleksandr Dugin, Putin's statements and actions make it abundantly clear that he is reading Dugin's hymnal and finding it useful. Based on this worldview, in February 2022 he ordered the invasion of Ukraine.

At this writing, Putin's invasion forces have been halted, and partially pushed back. With proper Western backing, Ukraine may be able repel the current invasion entirely. But wherever the ceasefire lines are drawn, more and bigger attacks are inevitable so long as the belief in a zero-sum world predominates.

If war is seen as necessary, then cults that sell war will not lack for sponsors, or recruits.

Hitler said that the idea of peace and plenty through scientific progress was a Jewish plot to undermine people's belief in the necessity for war. He was half right. It is certainly not a Jewish plot. But it does undermine people's belief in the necessity for war. That is why we need it to prevail.

Do we really face the threat of general war? There seems to be no reason for it, and in fact, there isn't. People all over the world today are actually living much better than they ever did before, at any time in human history. But the same was true in 1914. As then, and again in 1939, all it takes is the *belief* that there isn't enough to go around, that others are using too much, or threatening by their growth to do so in the future, to set the world ablaze.

If it is accepted that the future will be one of resource wars, there are men of action who are prepared to act accordingly.

There is no scientific foundation supporting these motives for conflict. On the contrary, it is precisely because of the freedom and affluence of the United States that American citizens have been able to invent most of the technologies that have allowed China, Russia, and so many other countries to lift themselves out of poverty. And should China (with a population five times ours) develop to the point where its per-capita rate of invention mirrors that of the United States—with 4 percent of the world's population producing 50 percent of the world's inventions—the entire human race would benefit enormously. Yet, that is not how people see it, or are being led to see it by those who should know better.

Rather, people are being bombarded on all sides with zero-sum-derived propaganda not only by those seeking trade wars, immigration bans, or preparations for resource wars, but by those who, portraying humanity as a horde of vermin endangering the natural order, wish to use Malthusian ideology as justification for suppressing freedom. Such arguments sometimes costume themselves as environmentalist, but that is deception. True environmentalism takes a humanist point of view, seeking practical solutions for real problems in order to enhance the environment for the benefit of human life in its broadest terms. It therefore welcomes technological progress. Antihuman Malthusianism, on the other hand, seeks to make use of instances of inadvertent human damage to nature as an ideological weapon on behalf of the age-old reactionary thesis that humans are nothing but pests whose aspirations need to be contained and suppressed by tyrannical overlords to preserve a divinely ordered stasis.

"The Earth has cancer, and the cancer is man," proclaims the elite Club of Rome in one of its manifestos. This mode of thinking has clear implications. One does not provide liberty to vermin. One does not seek to advance the cause of a cancer.

The real lesson of the last century's genocides is this: *We are not endangered by a lack of resources. We are endangered by those who believe there is a shortage of resources. We are not threatened by the existence of too many people. We are threatened by people who think there are too many people.*

Those who accept the inevitability of general war but think they can mitigate its consequences by suppressing nuclear power have it exactly wrong. Atomic weapons can be (and were) manufactured in large numbers without there being any nuclear power plants. Moreover, much more deadly weapons can be created at much lower cost with vastly simpler facilities making use of biotechnology. We now know how to read the genetic code, and are rapidly learning how to write it. This technology offers innumerable benefits, as it will make possible the engineering of microbes – self reproducing machines if you will – programmed to make all kinds of useful substances from foods and fabrics to pharmaceuticals. But it will also enable the creation and mass production of disease organisms designed to kill, or avoid killing, members of particular ethnic groups. Using such technology, any group or nation with a few hundred million dollars to spend that feels itself aggrieved and without hope could wreak havoc upon the world.

The existential threat facing humanity is not climate change. It is the ideologies of despair.

If the twenty-first century is to be one of peace, prosperity, hope, and freedom, a definitive and massively convincing refutation of these pernicious beliefs is called for—one that will forever tear down the walls of the hellish prison these ideas would create for humanity.

The most dramatic refutation possible would be to unleash the unlimited potential of nuclear power.

## A QUESTION OF FAITH

> *We believe that free labor, that free thought, have enslaved the forces of nature, and made them work for man. We make the old attraction of gravitation work for us; we make the lightning do our errands; we make steam hammer and fashion what we need.... The wand of progress touches the auction-block, the slave-pen, the whipping-post, and we see homes and firesides and schoolhouses and books, and where all was want and crime and cruelty and fear, we see the faces of the free.*[12]
>
> —Colonel Robert G. Ingersoll,
> Indianapolis speech, 1876

Western civilization is based on the radical individualist proposition advanced by the Greek philosophers Socrates and Plato that there is an innate faculty of the human mind capable of distinguishing right from wrong, justice from injustice, truth from untruth. Embraced by early Christianity, this idea became the basis of the concept of the *conscience*, which thereupon became the axiomatic foundation of Western morality. It is also the basis of our highest notions of law – the Natural Law determinable as justice by human conscience and reason, put forth, for example, in the US Declaration of Independence ("We hold these truths to be self-evident...") – from which we draw our belief in the fundamental rights of man existing independently of any laws that may or may not be on the books or existing accepted customs. It is also the basis for *science*, man's search for universal truth through the tools of reason.

As the great Renaissance scientist Johannes Kepler, the discoverer of the laws of planetary motion put it; "Geometry is one and eternal, a reflection out of the mind of God. That mankind shares in it is one reason to call man the image of God." In other words, the human mind, because it is the image of God, is able to understand the laws of the universe. It was the forceful demonstration of this proposition by Kepler, Galileo, and others that let loose the scientific revolution in the West.

Science, reason, morality based on individual conscience, human rights; this is the Western humanist heritage. Whether expressed in Hellenistic, Christian, Deist, or purely Naturalistic forms, it all drives toward the assertion of the fundamental dignity of man. As such, it rejects human sacrifice and is ultimately incompatible with slavery, tyranny, ignorance, superstition, perpetual misery, and all other forms of oppression and degradation. It asserts that humanity is capable and worthy of progress.

This last idea – progress – is the youngest and proudest child of Western humanism. Born in the Renaissance, it has been the central motivating idea of our society for the past four centuries. As a civilizational project to better the world for posterity, its results have been spectacular, advancing the human condition in the material, medical, legal, social, moral, and intellectual realms to an extent that has exceeded the wildest dreams of its early utopian champions.

Yet now it is under attack. It is being said that the whole episode has been nothing but an enormous mistake, that in liberating ourselves we have destroyed the Earth. As influential Malthusians Paul Ehrlich and John Holdren put it in their 1971 book *Global Ecology*: "When a population of organisms grows in a finite environment, sooner or later it will encounter a resource limit. This phenomenon, described by ecologists as reaching the 'carrying capacity' of the environment, applies to bacteria on a culture dish, to fruit flies in a jar of agar, and to buffalo on a prairie. It must also apply to man on this finite planet."

We need to refute this. The issue before the court is the fundamental nature of humankind. Are we destroyers or creators? Are we the enemies of life or the vanguard of life? Do we deserve to be free?

Ideas have consequences. Humanity today faces a choice between two very different sets of ideas, based on two very different visions of the future. On the one side stands the anti-human view, which, with complete disregard for its repeated prior refutations, continues to postulate a world of limited supplies, whose fixed constraints demand ever-tighter controls upon human aspirations. On the other side stand those who believe in the power of unfettered creativity to invent unbounded resources and so, rather than regret human freedom, demand it as our birthright. The contest between these two outlooks will determine our fate.

If the idea is accepted that the world's resources are fixed with only so much to go around, then each new life is unwelcome, each unregulated act or thought is a menace, every person is fundamentally the enemy of every other person, and each race or nation is the enemy of every other race or nation. The ultimate outcome of such a worldview can only be enforced stagnation, tyranny, war, and genocide. Only in a world of unlimited resources can all men be brothers.

On the other hand, if it is understood that unfettered creativity can open unbounded resources, then each new life is a gift, every race or nation is fundamentally the friend of every other race or nation, and the central purpose of government must not be to restrict human freedom, but to defend and enhance it at all costs.

It is for this reason that we need urgently to break the chains holding back nuclear power on Earth, create universal prosperity, and open vast new frontiers in space. We must joyfully embrace the challenge of launching new, dynamic, pioneering branches of human civilization on Mars—so that their optimistic, impossibility-defying spirit will continue to break barriers and point the way to the incredible plentitude of possibilities that urge us to write our daring, brilliant future among the vast reaches of the stars. We need to show for all to see in the most sensuous way possible what the great Italian Renaissance humanist Giordano Bruno boldly proclaimed; "there are no ends, limits, or walls that can bar us or ban us from the infinite multitude of things."

Bruno was burned at the stake by the Inquisition for his daring, but fortunately others stepped up to carry the banner of reason, freedom, and dignity forward to victory in his day. So we must do in ours.

Fig. 16.3 A statue of free thought hero Giordano Bruno stands today in Rome's Campo de' Fiori square where he was burned at the stake in 1600. There is no statue dedicated to those who burned him. Be like Bruno.

And that is why we must unleash the nuclear energy revolution. For in doing so, we make the most forceful statement possible that we are living not at the end of history, but at the beginning of history; that we believe in freedom and not regimentation, in progress and not stasis, in love rather than hate, in peace rather than war, in life rather than death, and in hope rather than despair.

# REFERENCES

## CHAPTER 1 TOO LITTLE FIRE, TOO MUCH SMOKE

1. "Measuring Atmospheric Greenhouse Gases: Behind the Scenes," Global Carbon Monitoring Laboratory, National Oceanic and Atmospheric Administration, https://gml.noaa.gov/outreach/behind_the_scenes/gases.html (Accessed October 30, 2021)
2. Adam Frank, *Light of the Stars: Alien Worlds and the Fate of the Earth*, (New York: W.W. Norton, 2018)
3. Leigh Burrows, "Deaths from Fossil Fuel Emissions Higher than Previously Thought: Fossil fuel air pollution responsible for more than 8 million people worldwide in 2018," *Harvard School of Engineering*, February 9, 2021, https://www.seas.harvard.edu/news/2021/02/deaths-fossil-fuel-emissions-higher-previously-thought (Accessed October 30, 2021.) See also Petr Beckmann, *The Health Hazards of Not Going Nuclear*, (Boulder, CO: Golem Press, 1976). Gwyneth Cravens, *Power to Save the World: The Truth About Nuclear Energy*, (New York: Vintage Books, 2007).

4. NASA, *Carbon Dioxide Fertilization Greening the Earth, Study Finds*, https://www.nasa.gov/feature/goddard/2016/carbon-dioxide-fertilization-greening-earth Accessed November 17, 2018

## CHAPTER 2 A BRIEF HISTORY OF POWER

1. V. Gordon Childe, *Man Makes Himself,* (New York: New American Library,1951)
2. "Antipater of Thessalonica," *Wikipedia* https://en.wikipedia.org/wiki/Antipater_of_Thessalonica (Accessed October 30, 2021)
3. Frances and Joseph Gies, Cathedral, Forge, and Waterwheel: Technology and Invention in the Middle Ages, (New York: Harper Collins, 1994).
4. Michael Shellenberger, *Apocalypse Naver: Why Environmental Alarmism Hurts Us All,* (New York: Harper Collins, 2020). Written by a noted ecomodernist, this book powerfully argues the theme that "to save the natural, we must embrace the artificial."
5. Data for total world Gross Domestic Product are estimates published Professor J. Bradford DeLong of the Department of Economics, U.C. Berkeley. See http://en.wikipedia.org/wiki/Gross_world_product (accessed October 30, 2021), and J. Bradford DeLong, "Estimating World GDP: 1 million BC to Present," https://delong.typepad.com/print/20061012_LRWGDP.pdf (accessed October 30, 2021). Population estimates are those compiled by Wikipedia from various experts, see articles on "World Population," http://en.wikipedia.org/wiki/World_population (accessed October 30, 2021) and "World Population Estimates," http://en.wikipedia.org/wiki/World_population_estimates (accessed October 30,

2021). Data on world energy use from the US Energy Information Administration, EIA.gov.

6.  Donella H. Meadows, Dennis Meadows, Jorgen Randers, and William W. Behrens, *The Limits to Growth: A Report for the Club of Rome's Project on the Predicament of Mankind,* (New York: Universe Publications, 1972)

7.  "U.S. Crude Oil and Natural Gas Proved Reserves, Year-end 2019, Energy Information Administration, https://www.eia.gov/naturalgas/crudeoilreserves/ (Accessed October 20, 2021)

8.  Ibid.

9.  Daniel Yergin, *The Prize: The Epic Quest for Oil, Money, and Power,* (New York: Simon and Schuster, 1991) Also see Robert Zubrin, *Energy Victory: Winning the War on Terror by Breaking Free of Oil* (Prometheus Books: Amherst, 2007)

10. Robert Zubrin, "Putin's Anti Fracking Campaign," *National Review*, May 5, 2014 https://www.nationalreview.com/2014/05/putins-anti-fracking-campaign-robert-zubrin/ (Accessed October 30. 2021)

11. Hobart King, "Methane Hydrate: The world's largest natural gas resource is trapped beneath permafrost and ocean sediments." *Geology.com* https://geology.com/articles/methane-hydrates/

12. W. Brian Arthur, *The Nature of Technology: What It Is and How It Evolves.* (New York: The Free Press, 2009)

13. William Fielding Ogburn, *Social Change with Respect to Culture and Ordinal Nature,* (New York: The Viking Press, 1922) Matt Ridley, *How Innovation Works: And Why It Flourishes in Freedom* (New York: Harper, 2020.)

## CHAPTER 3 DO WE REALLY NEED MORE ENERGY?

1.  Thomas Robert Malthus, An Essay on the Principle of Population, and Other Writings," (New York: Penguin, 2015) Original editions of Malthus were published between 1798 and 1826. Also see Allan Chase, *The Legacy of Malthus* (New York: Alfred A. Knopf, 1977) and Robert Zubrin, *Merchants of Despair: Radical Environmentalists, Criminal Pseudo-Scientists, and the Fatal Cult of Antihumanism*, (New York: Encounter Books, 2012.)

2.  In the recent movie, *The Boy with the Striped Pajamas*, the commandant of a Nazi concentration camp is confronted by his wife, who is upset by the mass murders going on in his facility. The commandant, who is no crude thug but a solid family man with clear managerial ability, tries to get her to understand the need for his work. "I'm just trying to make the world a better place," he tells her. This was indeed the view of many of those who devoted themselves to the Nazi cause, as well as most of those involved in the population control movement down to this day. They are just trying to make the world better by getting rid of undesirable people. The consequences are invariably the same.

3.  Data for total world Gross Domestic Product are estimates published Professor J. Bradford DeLong of the Department of Economics, U.C. Berkeley. See http://en.wikipedia.org/wiki/Gross_world_product (accessed October 30, 2021), and J. Bradford DeLong, "Estimating World GDP: 1 million BC to Present," https://delong.typepad.com/print/20061012_LRWGDP.pdf (accessed October 30, 2021). Population estimates are those compiled by Wikipedia from various experts, see articles on "World Population," http://en.wikipedia.org/wiki/World_population (accessed October 30, 2021) and

"World Population Estimates," http://en.wikipedia.org/wiki/World_population_estimates (accessed October 30, 2021)

4. Julian L. Simon, *The Ultimate Resource 2*, (Princeton, NJ: Princeton University Press, 1996). This is an update and substantial expansion of the original edition, *The Ultimate Resource*, published in 1981. You can read either, but the newer one is easier to get, and has excellent commentary discussing how the predictions in the first edition turned out (they were almost all correct), as well as refutations of arguments advanced against the first edition by Simon's critics. This book is the best systematic refutation of Malthusian theory written by anyone, ever. I recommend it strongly.

5. Data from sources cited in reference 3, above.

6. Matt Ridley, *How Innovation Works: And Why It Flourishes in Freedom* (New York: Harper, 2020.) Ray Kurzweil, *The Singularity is Near*, (New York: Viking Press, 2005)

7. "Abundance of Elements in the Earth's Crust," Wikipedia, https://en.wikipedia.org/wiki/Abundance_of_elements_in_Earth%27s_crust#cite_note-7 accessed August 16, 2020.

8. Frances and Joseph Gies, Cathedral, Forge, and Waterwheel: Technology and Invention in the Middle Ages, (New York: Harper Collins, 1994).

9. Lynn White, Jr., Medieval Technology and Social Change, (London: Oxford University Press, 1962)

10. Jean Gimpel, The Medieval Machine: The Industrial Revolution of the Middle Ages, (New York: Barnes and Noble, 1976)

## CHAPTER 4 WHAT IS NUCLEAR ENERGY

1.   Florian Cajori, "The Age of the Sun and the Earth: Some Ancient and Modern Theories," *Scientific American*, September 12, 1908 https://www.scientificamerican. com/article/the-age-of-the-sun-and-the-earth/ (Accessed October 30, 2021)
2.   "Stellar Nucleosynthesis," Wikipedia, https://en.wikipedia. org/wiki/Stellar_nucleosynthesis (Accessed October 30, 2021)
3.   Robert Zubrin, Energy Victory: Winning the War on Terror by Breaking Free of Oil (Amherst, NY: Prometheus Books, 2007), 210. Also see: Ronald Knief, Nuclear Energy Technology: Theory and Practice of Commercial Nuclear Power (New York: McGraw Hill, 1981), 549; George A. Olah, Alain Goeppert, and G. K. Surya Prakash, Beyond Oil and Gas: The Methanol Economy (Weinheim, Germany: Wiley-VCH, 2006), 27-50; BP, "BP Statistical Review of World Energy, June 2011," 2011, www. bp.com/statisticalreview; and U.S. Bureau of the Census, "International Data Base: Total Midyear Population for the World: 1950-2050," June 2011, http://www.census. gov/population/international/data/idb/worldpoptotal. php.

## CHAPTER 5 HOW NUCLEAR ENERGY CAME TO BE

1.   Richard Rhodes, *The Making of the Atomic Bomb*, (New York: Simon and Schuster, 1986)
2.   Table of Simple Integral Neutron Cross-Section Data from JEF-1, ENDF/B-V, ENDL-82, JENDL-2, KEDAK-4, RCN-3. NEA Data Bank, July 1985, https://www.oecd-nea.org/dbdata/nds_jefreports/jefreport-3.pdf (accessed October 30, 2021.)

3. For the basic theory of reactor neutronics, see John R. LaMarsh, *Introduction to Nuclear Engineering,* (New York: Addison Wesley, 1975)

4. David Irving, *The German Atomic Bomb.*(New York: Simon and Schuster, 1968)

5. J. Robert Oppenheimer, "Now I am Become Death" Atomic Archive, https://www.atomicarchive.com/media/videos/oppenheimer.html (Accessed October 30, 2021)

6. *Leona Woods*, Wikipedia, https://en.wikipedia.org/wiki/Leona_Woods (Accessed October 30, 2021)

## CHAPTER 6 ATOMS FOR SUBS, ATOMS FOR PEACE

1. Norman Polmar and Thomas B. Allen: *Rickover: Controversy and Genius*; (New York; Simon and Schuster, 1982)

2. E. Kintner, "Admiral Rickover's Gamble: The Landlocked submarine," *Atlantic Monthly,* January 1959, p.31.

3. Richard Hewlett and Francis Duncan, *Nuclear Navy: 1946-1962*, (Chicago: University of Chicago Press, 1974)

4. H. Rickover, private communication, 1980

5. "Eisenhower's 'Atoms for Peace' Speech." *Atomic Heritage* https://www.atomicheritage.org/key-documents/eisenhowers-atoms-peace-speech (Accessed October 31, 2021)

## CHAPTER 7 HOW TO BUILD A NUCLEAR REACTOR

1. Petr Beckmann, *The Health Hazards of Not Going Nuclear*, (Boulder, CO: Golem Press, 1976)

## CHAPTER 8 IS NUCLEAR POWER SAFE?

1.  U.S. Nuclear Regulatory Commission, "Fact Sheet on Biological Effects of Radiation," January 2011, last reviewed/updated February 4, 2011, http://www.nrc. gov/reading-rm/doc-collections/fact-sheets/bio-effects-radiation.html. See also, National Council on Radiation Protection and Measurements, "Uncertainties in Fatal Cancer Risk Estimates Used in Radiation Protection," NCRP Report, no. 126 (1997).
2.  Robert Zubrin, *Merchants of Despair: Radical Environmentalists, Criminal Pseudo-Scientists, and the Fatal Cult of Antihumanism*, (New York: Encounter Books, 2012.) pp. 62, 65, 86.
3.  Petr Beckmann, The Health Hazards of Not Going Nuclear, 113-114; Health Physics Society, "What Are Some Sources of Radiation Exposure?" updated August 27, 2011, http://hps.org/publicinformation/ ate/faqs/radsources.html (Accessed October 30, 2021); "Background Radiation" Wikipedia https://en.wikipedia. org/wiki/Background_radiation
4.  The Environmental Protection Agency website offers a calculator that lets you estimate your annual radiation dosage: https://www.epa.gov/radiation/calculate-your-radiation-dose (Accessed October 30, 2021)
5.  Richard Rhodes, "Radioactive Coal Ash," *New York Times*, September 8, 2010.
6.  Charles D. Hollister, D. Richard Anderson, and G. Ross Heath, "Subseabed Disposal of Nuclear Wastes," *Science* 213, no. 4514 (1981): 1321-1326. See also Scientia Press, "Sea-Based Nuclear Waste Solutions," https:// www.scientiapress.com/nuclearwaste (Accessed October 30, 2021)

7. "Yucca Mountain Science and Engineering Report" (U.S. Department of Energy, Rev. 1 to DOE/RW-0539, North Las Vegas, NV: U.S. DOE 2002), 4-463, https://www. nrc.gov/docs/ML0305/ML030580066.pdf (Accessed October 30, 2021)

8. "Yucca Mountain Waste Repository Development" Hearing before the Committee on Energy and Natural Resources, United States Senate, May 16, 2002 https:// www.govinfo.gov/content/pkg/CHRG-107shrg79940/ html/CHRG-107shrg79940.htm (Accessed October 30, 2021)

9. "Yucca Mountain: The Most Studied Real Estate on the Planet," Report to the Chairman, prepared by majority staff, Senate Committee on Environment and Public Works, March 2006. http://large. stanford.edu/courses/2015/ph241/avery-w1/docs/ YuccaMountainEPWReport.pdf (Accessed October 30, 2021)

10. Hannah Northey, "GAO: Death of Yucca Mountain Caused by Political Maneuvering," Greenwire for NYtimes.com, May 10, 2011, https://archive.nytimes. com/www.nytimes.com/gwire/2011/05/10/10greenwire-gao-death-of-yucca-mountain-caused-by-politica-36298. html?pagewanted=allzz (Accessed October 30, 2021)

11. "Democrats' Mission to Kill Yucca Mountain Rejects Sound Science, American Energy Security, and American Jobs," Senate Committee on Environment and Public Works, press release, May 7, 2009, https://www.legistorm. com/stormfeed/view_rss/18166/office/680/title/ democrats-mission-to-kill-yucca-mountain-rejects-sound-science-american-energy-security-and-american-jobs. html (Accessed October 30, 2021)

12. Christine Perham, "EPA's Role at Three Mile Island" EPA Journal 6, no. 9 (1980), available at https://archive.

epa.gov/epa/aboutepa/epas-role-three-mile-island.html (Accessed October 30, 2021).

13. Government of Japan, "Occurrence and Development of the Accident at the Fukushima Nuclear Power Stations" in Report of Japanese Government to IAEA Ministerial Conference on Nuclear Safety: Accident at TEPCO's Fukushima Nuclear Power Stations (Nuclear Emergency Response Headquarters, June 2011), https://japan.kantei.go.jp/kan/topics/201106/pdf/chapter_iv_all.pdf (Accessed October 30, 2021) .Robert Zubrin, "Fire NRC Chairman Gregory Jaczko," National Review Online, June 13, 2011, http://www.nationalreview.com/articles/269475/fire-nrc-chairman-gregory-jaczko-robert-zubrin. (Accessed October 30, 2021)

14. The Chernobyl Forum, Chernobyl's Legacy: Health, Environmental, and Socioeconomic Impacts (Vienna, Austria: IAEA, 2005), 16.

15. This is because commercial reactors keep their fuel in place for a long time, during which some of the 239Pu created in the reactor absorbs a further neutron to become 240Pu. The 240Pu seriously degrades the value of the plutonium for weapons purposes. However, in standalone atomic piles, such as those developed at the Hanford Site during the Manhattan Project, the fuel is not left in the system for long, so the plutonium produced is not spoiled. In the case of thorium reactors, which breed 232Th to 233U, the use of reactor fuel for bomb-making becomes even more difficult, making such systems ideal for use in situations where proliferation is of concern.

16. Amy Maxmen and Smiriti Malapaty, "The COVID lab-leak hypothesis: what scientists do and don't know," *Nature*, June 8, 2021 https://www.nature.com/articles/d41586-021-01529-3 (Accessed October 30, 2021)

## CHAPTER 9 HOW TO CUT COSTS

1. Bernard Cohen, *The Nuclear Energy Option, Chapter 9, The Costs of Nuclear Power Plants- What Went Wrong*, http://www.phyast.pitt.edu/~blc/book/chapter9.html (Accessed October 30, 2021)
2. ibid.
3. Jessica Lovering, Arthur Yip, and Ted Nordhaus, "Historical Construction Costs of Global Nuclear Power Reactors," Energy Policy, Volume 91, April 2016, Pages 371-382 https://www.sciencedirect.com/science/article/pii/S0301421516300106 (Accessed October 31, 2021)
4. Brad, Plumer, "Why America abandoned nuclear power (and what we can learn from South Korea)," *Vox*, Feb 29, 2016 https://www.vox.com/2016/2/29/11132930/nuclear-power-costs-us-france-korea (Accessed October 31, 2021)
5. "Shoreham Nuclear Power Plant," Wikipedia, https://en.wikipedia.org/wiki/Shoreham_Nuclear_Power_Plant (Accessed October 30, 2021)
6. Robert Zubrin, *Merchants of Despair*, (New York: Encounter Books, 2012) Figures are from Influence Watch, https://www.influencewatch.org/non-profit/world-wildlife-fund/ (Accessed Nov 1, 2021). They vary between 2017 and 2020 report. Sierra Club and Greenpeace figures combine the income and assets of the organizations with those of their foundations.
7. Robert Zubrin, *Merchants of Despair*, especially chapters 8,10, 11, 12, 13, and 14.
8. "Technology Roadmap Nuclear Energy 2015", IEA https://www.iea.org/reports/technology-roadmap-nuclear-energy-2015 (Accessed Nov 1, 2021)
9. Michael Shellenberger, "Why the War on Nuclear Threatens Us All," *Environmental Progress*, March

28, 2017 https://environmentalprogress.org/
big-news/2017/3/28/why-the-war-on-nuclear-threatens-
us-all (Accessed November 1, 2021)

10. Michael Shellenberger, *Apocalypse Never: How
    Environmental Alarmism Hurts Us All* (New York: Harper,
    2020). See especially pages 200-222.

## CHAPTER 10 BREEDING MORE FUEL THAN YOU BURN

1.  Alvin M. Weinberg, *The First Nuclear Era: The Life and
    Times of a Technological Fixer*, (New York: AIP Press,
    1994), pp. 40, 110. https://www.researchgate.net/
    publication/283847146_Fuel_cycle_performance_of_
    intermediate_spectrum_reactors_with_UTh_feed_and_
    continuous_recycling_of_UTRU_and_ThU3 (Accessed
    October 30, 2021)

2.  "LMFBR Schematics," *Wikipedia Commons*, https://
    commons.wikimedia.org/wiki/File:LMFBR_schematics.
    png (Accessed October 30, 2021)

3.  Alvin M. Weinberg, *The First Nuclear Era: The Life and
    Times of a Technological Fixer*, (New York: AIP Press,
    1994) pp.158

4.  "Experimental Breeder Reactor-2," *Wikipedia* https://
    en.wikipedia.org/wiki/Experimental_Breeder_Reactor_II
    (Accessed October 30, 2021) Also see Charles Till and
    Yoon Chang, *Plentiful Energy: The Story of the Integral Fast
    Reactor,* (Seattle: Createspace, 2011)

5.  "Breeder Reactor," *Wikipedia*, https://en.wikipedia.org/
    wiki/Breeder_reactor (Accessed October 30, 2021)

6.  Rod Adams, "Light Water Breeder Reactor: Adapting
    A Proven System," *Atomic Insights*, Oct 1, 1995 https://
    atomicinsights.com/light-water-breeder-reactor-adapting-
    proven-system/ (Accessed October 30, 2021)

7.   James A. Lane, "Aqueous Homogenous Reactors," Oak Ridge National Laboratory, 1958 https://moltensalt.org/references/static/downloads/pdf/FFR_chap01.pdf (Accessed October 30, 2021) L. G. Alexander D. A. Carrison H. G. MacPherson J. T. Roberts, Nuclear Characteristics of Spherical Homogenous Two-Region Molten-Fluoride Salt Reactors, Oak Ridge National Lab, 1959 https://www.osti.gov/servlets/purl/4200482 (Accessed October 30, 2021)

8.   Alvin M. Weinberg, *The First Nuclear Era: The Life and Times of a Technological Fixer*, (New York: AIP Press, 1994) pp.120-121

## CHAPTER 11 ENTREPRENEURIAL NUKES

1.   "Generation IV Reactor," *Wikipedia*, https://en.wikipedia.org/wiki/Generation_IV_reactor (Accessed October 30, 2021)

2.   Rod Walton, "Feds seeking comment on certifying NuScale SMR design for U.S. next-gen nuclear," *Power Engineering*, July 6, 2021 https://www.power-eng.com/nuclear/feds-seeking-comment-on-certifying-nuscale-smr-design-for-u-s-next-gen-nuclear/#gref (Accessed October 30, 2021)

3.   "Small Nuclear Power Reactors," *World Nuclear Association*, September 2021 https://www.world-nuclear.org/information-library/nuclear-fuel-cycle/nuclear-power-reactors/small-nuclear-power-reactors.aspx (Accessed October 30, 20210

4.   "What Should I Do if a Small Modular Reactor Loses Off-Site Power? " *Office of Nuclear Energy*, September 2020, https://www.energy.gov/ne/articles/

what-should-i-do-if-small-modular-reactor-loses-site-power (Accessed October 30, 2021)

5.  "High Temperature Gas Reactors," *Science Direct*, 2019 https://www.sciencedirect.com/topics/engineering/high-temperature-gas-reactors (Accessed October 30, 2021)

6.  "TRISO Particles: The Most Robust Nuclear Fuel on Earth," *Office of Nuclear Energy,* July 9, 2019. https://www.energy.gov/ne/articles/triso-particles-most-robust-nuclear-fuel-earth (Accessed October 30, 2021)

7.  "Very High Temperature Reactor VHTR" *Generation IV International Forum,* https://www.gen-4.org/gif/jcms/c_42153/very-high-temperature-reactor-vhtr (Accessed October 30, 2021)

8.  Georgy Toshinsky and Vladimir Petrochenko "Modular Lead-Bismuth Fast Reactors in Nuclear Power," *Sustainability* 2012, 4, 2293-2316; doi:10.3390/su4092293 September 18, 2012 https://www.mdpi.com/2071-1050/4/9/2293/pdf (Accessed October 30, 2021)

9.  "Superphenix," *Wikipedia* https://en.wikipedia.org/wiki/Superph%C3%A9nix (Accessed October 30, 2021)

10.  Alvin M. Weinberg, *The First Nuclear Era: The Life and Times of a Technological Fixer*, (New York: AIP Press, 1994) pp.127-131

11.  "Plans for New Reactors Worldwide," *World Nuclear Association,* https://world-nuclear.org/information-library/current-and-future-generation/plans-for-new-reactors-worldwide.aspx (Accessed December 25, 2022)

## CHAPTER 12 THE POWER THAT LIGHTS THE STARS

1.  Hans Fantel, "Science Taps Star Power for Unlimited Energy," Popular Mechanics 145, no. 4 (1976): 71.

2.  Robert W. Conn et al., "Lower Activation Materials and Magnetic Fusion Reactors," Nuclear Technology/Fusion 5, no. 3 (1984): 291-310.

3.  George Liu et al., "The Rate of Methanol Production on a Copper-Zinc Oxide Catalyst: The Dependence on the Feed Composition," Journal of Catalysis 90, no. 1 (1984): 139-146. See also Robert Zubrin, *Energy Victory: Winning the War on Terrorism by Breaking Free of Oil*, (Amherst New York: Prometheus Books, 2007.)

4.  Kevin Bonsor, "How Fusion Propulsion Will Work," *How Stuff Works* https://science.howstuffworks.com/fusion-propulsion.htm (Accessed October 30, 2021)

5.  European Fusion Development Agreement, "The Current Status of Fusion Research," 2010, http://www.efda.org/fusion_energy/fusion_research_today.htm.

6.  Robert W. Conn, "Magnetic Fusion Reactors," in Edward Teller, ed., Fusion: Magnetic Confinement, vol. 1, pt. B (New York: Academic Press, 1981), 194-398.

7.  Robert Zubrin, "A Deuterium-Tritium Ignition Ramp for an Advanced Fuel Field-reversed Configuration Reactor," Fusion Technology 9, no. 1 (1986): 97-100.

8.  In view of the extraordinary potential benefits of controlled fusion, it is amazing that the U.S. Department of Energy's annual budget for fusion research is limited to about $675 million per year, about 3 percent of NASA's current budget. When compared to the cost of a single space shuttle flight, about $1 billion each (some 130 were flown), this number is truly remarkable. By contrast, no funding has been made available to build the next machine to advance on the fusion accomplishments achieved by the Princeton Plasma Physics Lab TFTR, built circa 1980. The same pattern can be observed in the European, Russian, and Japanese programs, none of which has been funded to build any new fusion experimental machines

of consequence since the late 1980s. This is why, in contrast to the earlier period, there has been no advance in experimental achievement in the fusion field since 1995. This period of stagnation can be laid at the doorstep of the bureaucrats of the U.S. Department of Energy and their counterparts in the European, Japanese, and Russian programs, who decided in the mid-1980s to replace their (supposedly wasteful, but actually quite productive) competing national programs with a single international cooperative program, called ITER (for International Tokamak Experimental Reactor). Lacking competitive spur, the ITER consortium has not achieved anything–not even a single experiment since it was launched 36 years ago.

This must certainly please the Malthusians, who made their true feelings about fusion clear in 1989, when an (unfortunately false) report appeared in the media that two chemists had made a breakthrough enabling cheap, practical, "cold fusion" energy without any radioactive wastes. In response to this apparent windfall for mankind, leading environmentalists could only express horror. On April 19, 1989, author Jeremy Rifkin told the Los Angeles Times that cold fusion was "the worst thing that could happen to our planet." In the same article, John Holdren, former president Obama's science advisor, also expressed his dismay over the invention, saying "clean-burning, non-polluting, hydrogen-using bulldozers still could knock down trees or build housing developments on farmland." Holdren's mentor, Population Bomb author Paul Ehrlich, went further (as always), stating that "industrialized societies, so far, have not used power wisely"–and thus cold fusion, even if clean and cheap, would be "like giving a machine gun to an idiot child." Bjørn Lomborg, The Skeptical Environmentalist (Cambridge, UK: Cambridge University Press, 2001), 321.

9. Daniel Clery, *A Piece of the Sun: The Quest for Fusion Energy*, (New York: Abrams Press, 2013). Clery's book provides the best account of the early days of controlled fusion research.

10. "Thermonuclear Weapon," *Wikipedia,* https://en.wikipedia.org/wiki/Thermonuclear_weapon (Accessed October 30, 2021.) For an epic account of H-Bomb development, see Richard Rhodes, *Dark Sun: The Making of the Hydrogen Bomb,*" (New York: Simon and Schuster, 1995)

11. An excellent account of Thonemann's life and work can be found in Daniel Clery, *A Piece of the Sun: The Quest for Fusion Energy*, (New York: Abrams Press, 2013).

12. Robin Herman, *Fusion: The Search for Endless Energy*, (Cambridge: Cambridge University Press, 1990) pp. 49-51. Daniel Clery, *A Piece of the Sun, p. 53*

13. Robin Herman, *Fusion*, p.52

14. Robin Herman, *Fusion*, pp.54-59

15. Robin Herman, *Fusion*, p.93

16. S.V. Mirnov "Tokamak Evolution and View to Future," Nucl. Fusion 59 015001, 2019 https://iopscience.iop.org/article/10.1088/1741-4326/aaee92/pdf (Accessed October 30, 2021)

17. Stephen O. Dean, *Search for the Ultimate Energy Source: A History of the U.S. Fusion Program*, (New York: Springer, 2013.) A former AEC then DOE official, Dean founded Fusion Power Associates. His book presents detailed account of the program's political history. The tale of the US fusion leaders fighting retreat in the face of ITER is heartbreaking.

18. Eric Berger, Liftoff: *Elon Musk and the Desperate Early Days that Launched SpaceX*, (New York: William Morrow, 2021)

19. R. L. Miller, R. A. Krakowski, C. G. Bathke, C. Copenhaver, N. M. Schnurr, A. G. Engelhardt, T. J. Seed, and R. M.

Zubrin *Advanced Tokamak Reactors Based on the Spherical Torus (ATR/ST) Preliminary Design Considerations,* Los Alamos National Lab report LA-10740-MS UC-20d Issued: June 1986 https://inis.iaea.org/collection/ NCLCollectionStore/_Public/18/043/18043683.pdf?r=1 (Accessed October 30, 2021)

20. Robert W. Bussard, "Galactic Matter and Interstellar Flight," *Astronautica Acta,* 1960, Volume 6, Fasc. 4 https:// web.archive.org/web/20180417180616/http://www. askmar.com/Robert%20Bussard/Galactic%20Matter%20 and%20Interstellar%20Flight.pdf (Accessed October 30, 2021)

21. George H. Miley and S. Krupakar Murali, *Inertial Electrostatic Confinement (IEC) Fusion: Fundamentals and Applications,* (New York: Springer, 2014)

## CHAPTER 13 OPENING THE SPACE FRONTIER

1. "SNAP 10A," *Wikipedia,* https://en.wikipedia.org/wiki/ SNAP-10A (Accessed October 30, 2021). For the most comprehensive discussion of space nuclear electric power systems see David Buden, *Space Nuclear Radioisotope Systems* (Lakewood, CO: Polaris Books, 2011) and David Buden *Space Nuclear Fission Electric Power Systems,* (Lakewood, CO: Polaris Books, 2011)

2. Sven Grahn, "The US-A program (Radar Ocean Reconnaissance Satellites – RORSAT) and radio observations thereof," *Space Tracking Notes,* http://www.svengrahn. pp.se/trackind/RORSAT/RORSAT.html (Accessed October 30, 2021)

3. "TOPAZ Nuclear Reactors," *Wikipedia* https://en.wikipedia. org/wiki/TOPAZ_nuclear_reactor (Accessed October 30, 2021)

4. "Kilopower," *Wikipedia*, https://en.wikipedia.org/wiki/Kilopower

5. David Buden, *Nuclear Thermal Propulsion Systems* (Lakewood, CO: Polaris Books, 2011) Also See Stanley K. Borowski, "Nuclear Thermal Propulsion," *Encyclopedia of Aerospace Engineering*, Wiley Online Library, December 2010 https://onlinelibrary.wiley.com/doi/abs/10.1002/9780470686652.eae115 (Accessed October 30, 2021)

6. H. Ludewig, J.R. Powell, M. Todosow, G. Maise, R. Barletta, D.G. Schweitzer "Design of particle bed reactors for the space nuclear thermal propulsion program," *Progress in Nuclear Energy,* Volume 30, Issue 1, 1996, Pages 1-65 https://www.sciencedirect.com/science/article/abs/pii/0149197095000804 (Accessed October 30, 2021)

7. Robert Zubrin with Richard Wagner, *The Case for Mars: The Plan to Settle the Red Planet and Why We Must,* (New York: Simon and Schuster, 1996)

8. Robert Zubrin, "The Profound Potential of Elon Musk's New Rocket," *Nautilus* Magazine, May 12, 2021 https://nautil.us/issue/100/outsiders/the-profound-potential-of-elon-musks-new-rocket (Accessed October 30, 2021)

9. . R.G. Ragsdale, "To Mars is 30 Days by Gas Core Nuclear Rockets," Astronautics and Aeronautics, 65, January 1972.

10. R.G. Ragsdale, "High specific Impulse Gas Core Reactors," NASA TM X-2243, NASA Lewis Research Center, March 1971.

11. T. Latham and C. Joyner, "Summary of Nuclear Light Bulb Development Status," AIAA 91-3512, AIAA/NASA/OAI Conference on Advanced SEI Technologies, Cleveland, Ohio, Sept. 1991.

12. A. Martin and A. Bond, "Nuclear Pulse Propulsion: A Historical Review of an Advanced Propulsion Concept,"

Journal of the British Interplanetary Society, Vol 32, pp 283-310, 1979.

13. R. Zubrin, "Nuclear Salt Water Rockets: High Thrust at 10,000 sec Isp," Journal of the British Interplanetary Society, Vol 44, pp 371-376, 1991.

14. S. K. Borowski, "A Comparison of Fusion/Antiproton Propulsion systems for Interplanetary travel," AIAA-87-1814, 23ed AIAA/ASME Joint Propulsion Conference, San Diego, CA, June 29-July 2, 1987.

15. John Nuckolls, Lowell Wood, Albert Thiessen, and George Zimmerman, "Laser Compression of Matter to Super-High Densities: Thermonuclear (CTR) Applications" *Nature* volume 239, pages139–142 (1972) https://www.nature.com/articles/239139a0 (Accessed December 16, 2022)

16. John H. Nuckolls, "Contributions to the Genesis and Progress of ICF" Inertial Confinement Nuclear Fusion: A historical Approach by its Pioneers. Edited by Guillermo Velarde and Natividad Santamarfa, (London: Foxwell & Davies, 2007) https://blog.nuclearsecrecy.com/wp-content/uploads/2021/05/01-Nuckolls-Contribs-Gen-Progress-ICF.pdf (Accessed December 16, 2022)

## CHAPTER 14 UPGRADING THE EARTH

1. EPA, *Climate Change Indicators: Length of Growing Season* https://www.epa.gov/climate-indicators/climate-change-indicators-length-growing-season Accessed October 30, 2021.

2. EPA, *Climate Change Indicators, US and Global Precipitation* https://www.epa.gov/climate-indicators/climate-change-indicators-us-and-global-precipitation Accessed October 30, 2021.

3. NASA, *Carbon Dioxide Fertilization Greening the Earth, Study Finds,* https://www.nasa.gov/feature/goddard/2016/carbon-dioxide-fertilization-greening-earth Accessed October 30, 2021

4. Russ George, "We can bring back healthy fish in abundance almost everywhere."

5. http://russgeorge.net/2014/04/11/bring-back-fish-everywhere/ Accessed October 30, 2021

6. NASA Goddard Spaceflight Center, http://disc.sci.gsfc.nasa.gov/giovanni/giovanni_user_images#iron_bloom_northPac. Accessed, November 17, 2018.

7. Naomi Klein, *This Changes Everything, Capitalism Against the Climate.* New York: Simon and Schuster, 2014.

8. Assuming 1 percent efficiency of phytoplankton in converting $CO_2$ into biomass, and an average day-night sunlight flux of 200 W per square meter, it would take 10 million square kilometers, or 3 percent of the oceans' total area, to capture all the $CO_2$ emissions from humanity's current 20 TW fossil fuel combustion. Or, put another way, humanity currently burns 10 billion tons of carbon per year while the ocean captures 100 billion tons. So a net increase of ten percent of total ocean productivity, which can be accomplished by raising a much smaller fraction of the oceans' desert expanses to match the productivity of its most fertile regions, would suffice to capture all human $CO_2$ emissions. If done in optimal fashion, up to 1 billion tons of food could be produced in the process, enough to provide about a pound per day to every person on Earth.

## CHAPTER 15 THE WAY FORWARD

1. Dan Bosch, "Opportunities for NEPA Reform at the NRC," *American Action Forum,* https://www.

americanactionforum.org/insight/opportunities-for-nepa-reform-at-the-nrc/#ixzz7AogOajFx

2. "Shoreham Nuclear Power Plant," Wikipedia, https://en.wikipedia.org/wiki/Shoreham_Nuclear_Power_Plant (Accessed October 30, 2021)

3. Michael Shellenberger, *Apocalypse Never: Why Environmental Alarmism Hurts Us All*, (New York: Harper Collins, 2020), chapter 10.

4. "How to Reform Nuclear Policy," *Clear Path*, https://clearpath.org/policy/nuclear/ (Accessed October 30, 2021)

5. Gwyneth Cravens, *Power to Save the World: The Truth About Nuclear Energy*, (New York: Vintage Books, 2007)

6. Cravens, part 5.

7. "The Breakthrough Institute," https://thebreakthrough.org/ (Accessed October 30, 2021) Also see "Breakthrough Institute," *Wikipedia* https://en.wikipedia.org/wiki/Breakthrough_Institute (Accessed October 30, 2021)

8. "Third Way". https://www.thirdway.org/ Accessed October 30, 2021. Also see "Third Way," Wikipedia https://en.wikipedia.org/wiki/Third_Way (Accessed October 30, 2021)

9. "An Ecomodernist Manifesto," http://www.ecomodernism.org/ (Accessed October 30, 2021)

10. J Todd Moss and Jessica Lovering, "Is Nuclear Power Pro Development? Six Reasons why the DFC Should Lift Its Ban," *Energy for Growth*, June 10, 2020

11. Jessica Lovering "Is Nuclear Power Only for Wealthy Countries?" *Energy for Growth*, July 20, 2020 https://www.energyforgrowth.org/blog/is-nuclear-power-only-for-wealthy-countries/ (Accessed October 30, 2021)

12. Jessica Lovering, Todd Moss, Jacob Kincer, Jackie Kempfer, and Josh Freed, "Mapping the Global Market for Advanced Nuclear," *Energy for Growth*, October 7,

2020 https://www.energyforgrowth.org/report/mapping-the-global-market-for-advanced-nuclear/ (Accessed October 30, 2021)

13. "Energy for Growth Hub," https://www.energyforgrowth.org/ (Accessed October 30, 2021)

14. "Good Energy Collective: A Progressive Policy Agenda for Advanced Nuclear," https://www.goodenergycollective.org/ (Accessed October 30, 2021) https://www.vox.com/energy-and-environment/2020/7/21/21328053/climate-change-nuclear-power-environmental-justice-energy-collective David Roberts, "Nuclear power has been top-down and hierarchical. These women want to change that." *Vox*, July 21, 2021 (Accessed Nov 1, 2021)

15. T Rauli Partanen and Janne M. Korhonen, *The Dark Horse: Nuclear Power and Climate Change,* (Helsinki: The National Library of Finland, 2020) Kirsty Gogan, Daniel Agerter, and Robert Stone, *Energy for Humanity*, https://energyforhumanity.org/wp-content/uploads/2020/11/EFH-ourstory-2020a-1.pdf (Accessed October 31, 2021.)

16. Weinberg, *The First Nuclear Era*, p.173

17. Robert Zubrin, "The Population Control Holocaust," *The New Atlantis*, Spring 2012. https://www.thenewatlantis.com/publications/the-population-control-holocaust (Accessed October 30, 2021)

## CHAPTER 16 THEIR PROGRAM AND OURS

1. Robert Zubrin, *Merchants of Despair*, New York: New Atlantis Books, 2012

2. Friedrich von Bernhardi, *Germany and the Next War*, translated by Allen H. Powles. New York: Longmans, Green & Co, 1914. German edition published in 1912.

3. Daniel Goldhagen, *Hitler's Willing Executioners*, New York: Vintage, 2007

4. Lizzie Collingham, *The Taste of War: World War II and the Battle for Food*, New York: Penguin, 2012.

5. Robert Zubrin, *Merchants of Despair*, Chapter 6.

6. Timothy Snyder, "The Next Genocide," *New York Times*, September 12, 2015, https://www.nytimes.com/2015/09/13/opinion/sunday/the-next-genocide.html Accessed October 31, 2021.

7. Timothy Snyder, *Black Earth: The Holocaust as History and Warning*, New York: Tim Duggan Books, 2015.

8. Michael Klare, *Resources Wars: The New Landscape of Global Conflict*. New York: Macmillan, 2001.

9. Robert Zubrin, Energy Victory: Winning the War on Terror by Breaking Free of Oil. Amherst, NY: Prometheus Books, 2007.

10. Robert Zubrin, "The Eurasianist Threat," *National Review*, March 3, 2014. https://www.nationalreview.com/2014/03/eurasianist-threat-robert-zubrin/ Accessed October 31, 2021. See also Robert Zubrin "The Wrong Right," *National Review*, June 24, 2014 https://www.nationalreview.com/2014/06/wrong-right-robert-zubrin/ Accessed December 28, 2022. Robert Zubrin, "Putin's Mad Philosopher," *National Review*, June 18, 2014 https://www.nationalreview.com/2014/06/dugins-evil-theology-robert-zubrin/ Accessed December 28, 2022. Robert Zubrin, "Putin's Rasputin: Meet Aleksandr Dugin, the Mystical High Priest of Russian Fascism Who Wants to Bring About the End of the World," *Skeptic Magazine*, March 26, 2022, updated from article published in *Skeptic Magazine* February 20, 2015. https://www.skeptic.com/reading_room/meet-aleksandr-dugin-mystical-high-priest-of-russian-fascism-who-wants-to-bring-about-end-of-the-world/ Accessed December 28, 2022.

11.  Robert Zubrin, "America Stop Breathing," *National Review* October 31, 2013. https://www.nationalreview.com/2013/10/america-stop-breathing-robert-zubrin/ Accessed October 31, 2021
12.  Robert G. Ingersoll, "Indianapolis speech, 1876: Delivered to the Veteran Soldiers of the Rebellion." https://infidels.org/library/historical/robert_ingersoll/indianapolis_speech76.html Accessed October 31, 2021

# ACKNOWLEDGEMENTS

I **WOULD LIKE** to acknowledge the help of Marie Stirk, book design specialist extraordinaire, for her fabulous job in the design and layout of the cover and text of this book.

The sources of information for this book are far too numerous to mention, but as I now near the close of my career, I would like to thank all those who made all my technical accomplishments, including this book, possible.

These started with a number of outstanding teachers. notably including my 5th and 6th grade teacher Mr. Allen, and my 8th grade science teacher Mr. Kolender. The former was a former US Air Force pilot. The latter had navigated for the Luftwaffe, before he was shot down in Italy, taken to America, to ultimately serve a far nobler cause. Then there was Mr. Miller, a serious scientist who taught 10th grade biology, and Mrs. Fox, a truly terrific teacher of 11th grade chemistry. All could have made a lot more money doing other things, but they chose instead to devote their lives to spreading their passion for science. I hope I have helped to pay them back.

I don't remember my undergraduate university professors as well, probably because that setting is much less intimate than those of either lower or higher educational levels. But in graduate school at the University of Washington, I came to know a number of professors who changed my life. These included Reiner Decher, Adam Bruckner, and Abe Hertzberg in the Aerospace Engineering Department, and, most of all, Fred Ribe, of the Department of Nuclear Engineering.

Fred Ribe, my graduate school mentor and thesis advisor, had been head of the fusion program at Los Alamos. His standards were very tough. He drove me up the wall. I owe him everything.

At this writing, Decher and Bruckner are retired. The rest have passed to dwell among the shadows. Yet they are not gone. The good never die.

# INDEX

# M

# N

# ABOUT THE AUTHOR

**DR. ROBERT ZUBRIN** is an internationally renowned nuclear and aerospace engineer with four decades of technical experience. Formerly a Senior Engineer at Lockheed Martin, since 1996 he has been President of Pioneer Astronautics, an aerospace research and development company. In that capacity he has led over 70 highly successful technology development projects for NASA, the US military, the Department of Energy, and private clients. He holds Master of Science degrees in Nuclear Engineering and Aeronautics and Astronautics, and a doctorate in Nuclear Engineering, all from the University of Washington. He is the author of 13 books, over 200 technical and non-technical papers in areas relating to aerospace and energy engineering, and is the inventor of over 20 US patents, with several more pending. In 1998 he founded the non-profit Mars Society, and personally led it in building a simulated human Mars exploration station in

the Canadian Arctic, some 900 miles from the North Pole. He remains president of the Mars Society today. Prior to his work in aerospace, Dr. Zubrin worked in areas of radiation protection, nuclear power plant safety, thermonuclear fusion research, and as a secondary school science and math teacher. He lives in Golden, Colorado with his wife Hope Zubrin, a retired Middle School science teacher. They have three daughters, Sarah, Rachel, and Oakley, all now out of the house, and a loyal Sheltie named Strelka and Siberian cat Luna, who remain at home.

CPSIA information can be obtained
at www.ICGtesting.com
Printed in the USA
BVHW041648110423
662157BV00005B/63